Klaus Lamotke

Regular Solids and Isolated Singularities

V

Friedr. Vieweg & Sohn Braunschweig/Wiesbaden

CIP-Kurztitelaufnahme der Deutschen Bibliothek

Lamotke, Klaus:
Regular solids and isolated singularities / Klaus
Lamotke. – Braunschweig; Wiesbaden: Vieweg,
1986.
(Advanced lectures in mathematics)
ISBN 3-528-08958-X

AMS Subject Classification: 14 J 17, 20 G 20, 32 B 30, 32 C 45, 51 F 25, 51 M 20, 57 MX.

1986

Produced by Lengericher Handelsdruckerei, Lengerich
Printed in Germany

ISBN 3-528-08958-X

Klaus Lamotke

Regular Solids and Isolated Singularities

Advanced Lectures in Mathematics

Edited by Gerd Fischer

Jochen Werner
Optimization. Theory and Applications

Manfred Denker
Asymptotic Distribution Theory
in Nonparametric Statistics

Klaus Lamotke
Regular Solids and
Isolated Singularities

Preface

The last book XIII of Euclid's Elements deals with the regular solids which therefore are sometimes considered as crown of classical geometry. More than two thousand years later around 1850 Schläfli extended the classification of regular solids to four and more dimensions. A few decades later, thanks to the invention of group and invariant theory the old three-dimensional regular solid were involved in the development of new mathematical ideas: F. Klein (Lectures on the Icosahedron and the Resolution of Equations of Degree Five, 1884) emphasized the relation of the regular solids to the finite rotation groups. He introduced complex coordinates and by means of invariant theory associated polynomial equations with these groups. These equations in turn describe isolated singularities of complex surfaces. The structure of the singularities is investigated by methods of commutative algebra, algebraic and complex analytic geometry, differential and algebraic topology. A paper by DuVal from 1934 (see the References), in which resolutions play an important rôle, marked an early stage of these investigations. Around 1970 Klein's polynomials were again related to new mathematical ideas: V.I. Arnold established a hierarchy of critical points of functions in several variables according to growing complexity. In this hierarchy Klein's polynomials describe the "simple" critical points.

The present book grew out of a two semester course at the University of Cologne for students with some basic knowledge of algebra, complex analysis, and topology, who wanted or needed another special course, mostly as their last encounter with Pure Mathematics. The book still reflects this situation: It presents basic material from commutative algebra to differential topology with some emphasis on several complex variables as well as the special topics from regular solids and isolated singularities mentioned above. As each chapter has its own introduction, no further review of the contents will be given here.

It is a pleasure to acknowledge the help of D. Wemmer who read the whole manuscript, and besides pointing out many small errors made valuable suggestions in order to make the text better understandable for the non-expert. My appreciation goes to Frau Becker who typed the various successive versions of the manuscript with great patience and skill.

Cologne, August 1985 K. Lamotke

TABLE OF CONTENTS

CHAPTER I

REGULAR SOLIDS AND FINITE ROTATION GROUPS

Five polyhedra in ordinary three-dimensional space have attracted special interest since ancient times because of their regularity: the tetrahedron, the cube (hexahedron), the octahedron, the icosahedron, and the dodecahedron. Platon describes them in the dialogue "Timaios" and thus they were called Platonic solids. We present them quite informally in §1. In order to answer the question why there are not more regular solids, the general concepts of convex polytopes and their regularity are introduced. We do this for arbitrary finite dimensions and show that in all dimension > 2 the possibilities for regular solids are very restricted. From dimension five on only the analogs of the tetrahedron, the cube, the octahedron survive. Further study of the six possibilities in four dimensions is postponed to chapter II, §3. In this chapter I we restrict attention again to three dimensions from §5 on.

A close connection between the regular solids and the finite groups of rotations is established. The geometry of the regular solids, which can be felt and seen when you turn the solid in your hand, yields the more abstractly formulated properties of the groups, e.g. their presentation by generators and relations.

The almost classical reference for the contents of this chapter and the history of the subject is Coxeter (1). Our presentation differs from his and is more straight forward because we restrict attention to *convex* polytopes. Our classification of the finite subgroups of SO(3) using standard results from linear algebra rather than from spherical geometry is due to Weyl.

§1 The Platonic Solids

We begin with a pyramid based on an equilateral triangle. All its faces are triangles. We adjust the altitude so as to make all triangles equilateral and obtain the *tetrahedron*. Alternatively we may begin with a cube: There are two tetrahedra whose edges are the diagonals of the faces of the cube. Together they form the *stella octangula* of Johann Kepler. If we take the cube with vertices $(\pm 1, \pm 1, \pm 1)$ in $\mathbb{R}^3$ one tetrahedron of the stella octangula is the convex hull of the four vertices

$$(1,-1,1)\ ,\ (1,1,-1)\ ,\ (-1,-1,-1)\ ,\ (-1,1,1)$$

and the other one has the four antipodal vertices

$$(-1,1,-1)\ ,\ (-1,-1,1)\ ,\ (1,1,1)\ ,\ (1,-1,-1)\ .$$

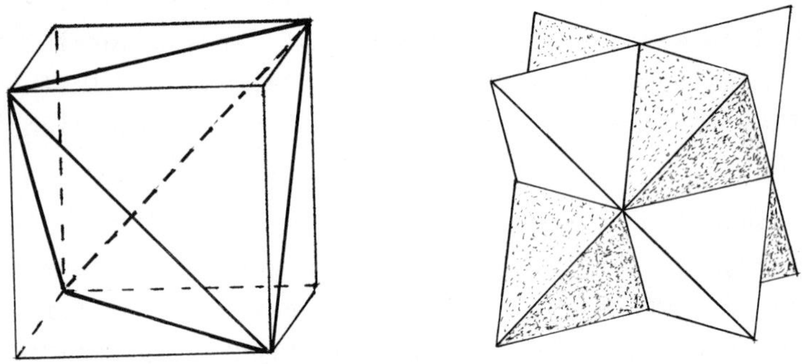

Next we take two pyramids based on congruent squares and adjust their altitudes so as to obtain equilateral triangles as faces. By placing the pyramids base to base we obtain the *octahedron*. Any convex solid has a *dual* namely the convex hull of the centres of its faces. The dual of a tetrahedron is again a (smaller) tetrahedron. The octahedron can also be obtained as the dual of a cube. (And the dual of an octahedron is again a smaller cube.) From the cube with vertices $(\pm 1, \pm 1, \pm 1)$ in $\mathbb{R}^3$ we obtain the dual octahedron with vertices $(\pm 1,0,0)$, $(0,\pm 1,0)$, and $(0,0,\pm 1)$.

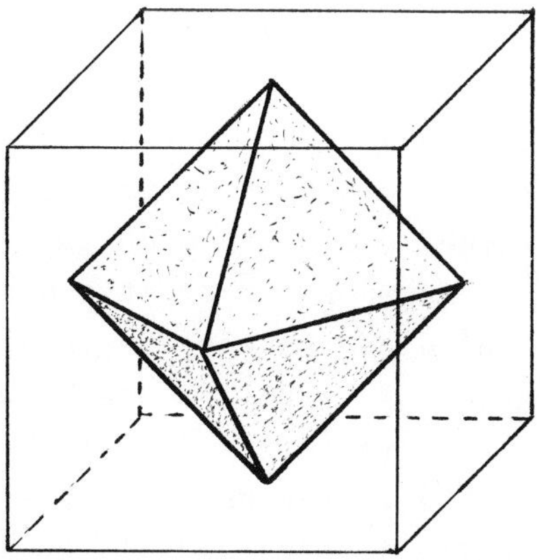

In order to construct the *icosahedron* we begin with a right prism based on a regular pentagon but look only at its bottom and top face (the two bases). We turn one of them around 36° in its own plane and form the convex hull of the twisted bases. This yields an antiprism with ten isoceles triangles as lateral faces whose ten lateral edges make a kind of zigzag. The altitude of the antiprism can be adjusted so as to make the lateral triangles equilateral. We then place right pyramids with equilateral triangles as lateral faces on the two bases of the antiprism and obtain the *icosahedron*.

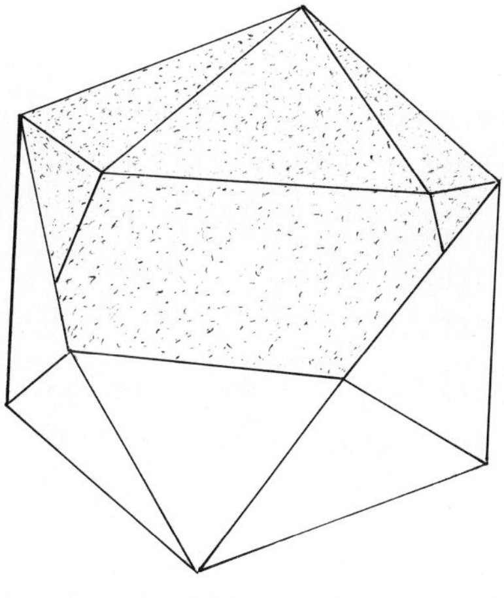

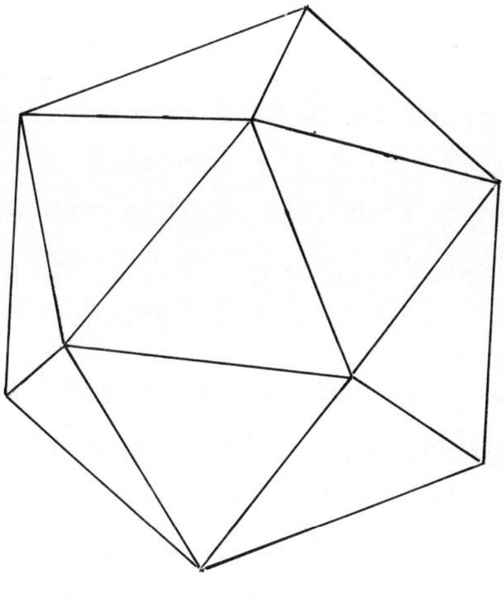

Another construction of the icosahedron begins with the octahedron: Its eight faces can be coloured alternately white and black like a chess board. (This is impossible when the number of faces with a common vertex is odd.)
The colouring determines a direction of each edge, so that there will be a white face on our right and a black one on our left if we proceed along an edge in the indicated direction. This enables us to define unambiguously, for any given ratio $a:b$, twelve points dividing the respective edges in this ratio. The convex hull of these twelve points is an icosahedron, which in general will be irregular: Its faces are twenty triangles. Eight of them lie on the faces of the octahedron. They are equilateral with sides $\sqrt{a^2+b^2-ab}$, if $a+b$ is the length of the edge of the octahedron. The other twelve triangles are isoceles with base $\sqrt{2}a$ and sides $\sqrt{a^2+b^2-ab}$.

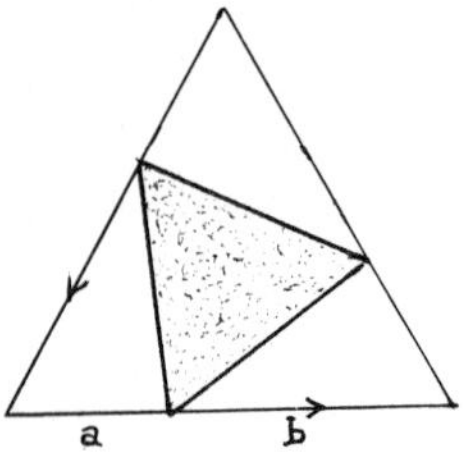

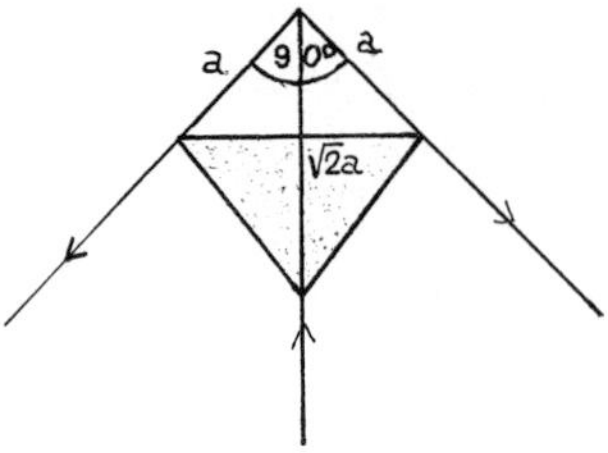

In order to obtain a regular icosahedron we choose the ratio $a:b$ so that the isoceles triangles become equilateral, i.e. $\sqrt{2}a = \sqrt{a^2+b^2-ab}$. This yields

$$a^2-b^2+ab = 0 \quad \text{or} \quad b:(a+b) = a:b \,, \quad \text{or}$$
$$a/b = \tau = \tfrac{1}{2}(\sqrt{5}-1) = 0{,}618033989\ldots \,.$$

Thus *the twelve vertices of the icosahedron are obtained by dividing the twelve edges of an octahedron according to the golden section.*

The octahedron with vertices $(\pm1,0,0)$, $(0,\pm1,0)$ and $(0,0,\pm1)$ has edge $\sqrt{2}$. If $a+b=1$ the edge joining $(0,0,1)$ and $(0,1,0)$ is divided in the ration $a:b$ by the point $(0,a,b)$. Such points on all the edges are the twelve points

$$(0,\pm a,\pm b) \ , \quad (\pm b,0,\pm a) \ , \quad (\pm a,\pm b,0) \ .$$

They are the vertices of a (regular) icosahedron if $a=1-\tau$ and $b=\tau$. Its edge has length $2(1-\tau)$.

Finally the dodecahedron is the dual solids of the icosahedron.

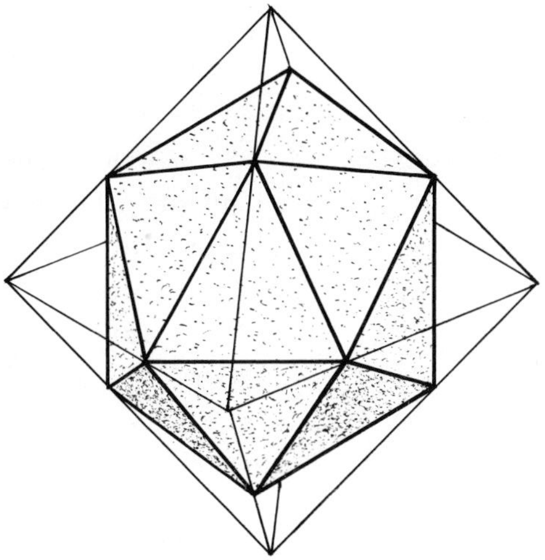

The vertices of the icosahedron divide the edges of the octahedron according to the golden section.

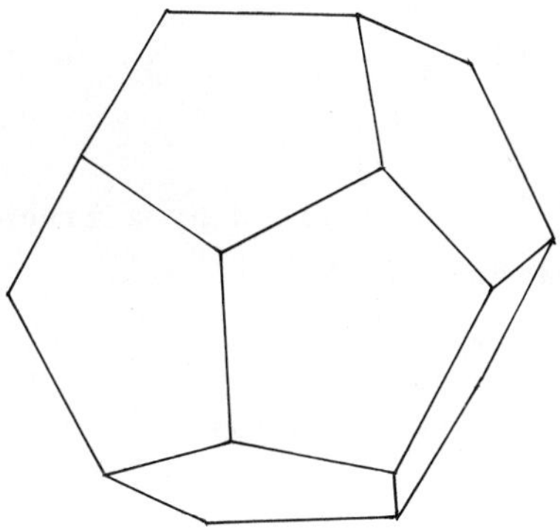

The dodecahedron

§2 Convex Polytopes

Two dimensional polytopes are polygons, three-dimensional polytopes are polyhedra. This section is a report on d-dimensional polytopes, which are convex. For more information the books by Grünbaum and Brøndsted are recommended.

We consider the Euclidean n-space $\mathbb{R}^n$ with the inner product $\langle -,- \rangle$ and the norm $|v| = \sqrt{\langle v,v \rangle}$. A map $f: \mathbb{R}^n \to \mathbb{R}^n$ is called a *Euclidean motion* if $|f(x)-f(y)|=|x-y|$ for any two $x,y \in \mathbb{R}^n$. The orthogonal transformations of $\mathbb{R}^n$ are exactly the Euclidean motions which have 0 as fixed point. The map $f: \mathbb{R}^n \to \mathbb{R}^n$ is called a $1:\varrho$ *similarity* for some $\varrho > 0$ if $x \mapsto f(x)/\varrho$ or equivalently $x \mapsto f(x/\varrho)$ is a Euclidean motion. Let $A,B \subset \mathbb{R}^n$ be two subsets with $f(A)=B$. If f is a Euclidean motion, A and B are called *congruent*; if f is a similarity, they are called *similar*. Affine subspaces of codimension 1 (hyperplanes) will be called *primes*.

Let M be a finite subset of $\mathbb{R}^n$. The convex hull

$$P = \operatorname{ch} M = \{ \sum_{m \in M} \lambda_m m \in \mathbb{R}^n : \lambda_m \geq 0 \text{ and } \Sigma \lambda_m = 1 \}$$

is called a *convex polytope*. There is a unique minimal subset $M^* \subset M$ such that $P = \operatorname{ch} M^*$. The elements of M^* are called the *vertices* of P. The *dimension* d of P is the dimension of the affine subspace spanned by M and hence by P.

A closed half space H of $\mathbb{R}^n$ is determined by a linear function $h: \mathbb{R}^n \to \mathbb{R}$ and a real number γ

$$H = \{ x \in \mathbb{R}^n : h(x) \leq \gamma \} .$$

The boundary of H is by definition the prime

$$\partial H = \{ x \in H : h(x) = \gamma \} .$$

The half space H contains the finite set M if and only if it contains the polytope $P = \operatorname{ch} M$. In this case

$$\partial H \cap P = ch(\partial H \cap M)$$

is called a *face* of P , see fig. 1. It is a (possibly empty)

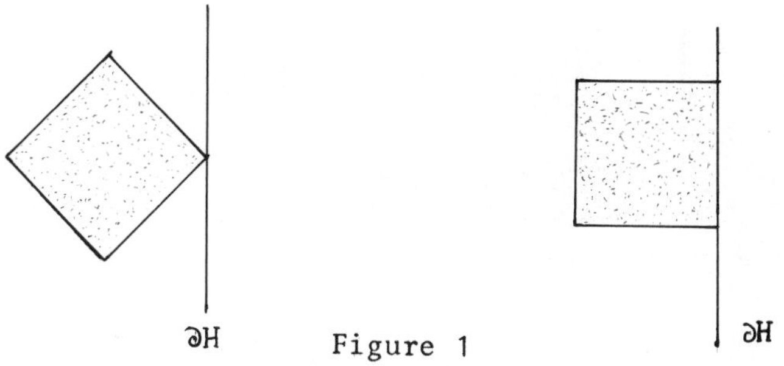

Figure 1

convex polytope. The whole polytope P is considered as improper face of itself. The zero-dimensional faces consist of one vertex each. The one-dimensional faces are called *edges*, and the (d-1)-dimensional faces of a d-dimensional polytope are called *facets*.

The number of faces of the Platonic solids

	tetrahedron	octahedron	cube	icosahedron	dodecahedron
vertices	4	6	8	12	20
edges	6	12	12	30	30
facets	4	8	6	20	12

If F is a face of P and G is a face of F , then G is a face of P . A sequence $P_o, P_1, \ldots, P_k$ of faces of P is called a *flag* of faces if each P_i is a face of P_{i+1} . The flag is *complete* if $\dim P_i = i$ and $k = \dim P$. Every flag is contained in a complete flag.

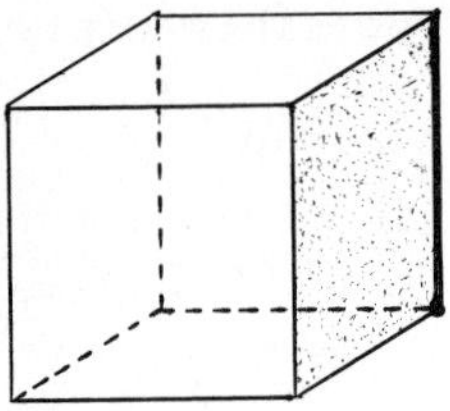

A flag of the cube

Figure 2

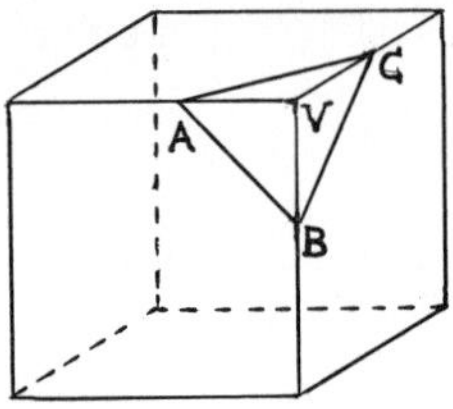

At the vertex V the triangle ABC is a vertex figure of the cube.

Figure 3

Let v be a vertex of P, and let H be a half space which contains all vertices of P but v. Then $Q = P \cap \partial H$ is called a *vertex figure* of P, see figure 3. The vertex figure Q is a convex polytope with the following property:

For every k*-face* F *of* P *which contains* v *the intersection* $F \cap \partial H$ *is a* (k-1)*-face of* Q. *This* $F \cap \partial H$ *is a vertex figure of* F *at* v. *The map* $F \to F \cap \partial H$ *establishes a* 1-1 *correspondence between the faces of* P *containing* v *and all faces of* Q. *The correspondence preserves flags.*

Let 0 be an interior point of the d-dimensional convex polytope $P \subset \mathbb{R}^d$. The *polar* P_r^o of P with respect to the (d-1)-sphere of radius r about 0,

$$P_r^o = \{x \in \mathbb{R}^d : \langle x,y\rangle \le r^2 \text{ for all } y \in P\}$$

is also a d-dimensional convex polytope. The double polar is $(P_r^o)_r^o = P$. If F is a k-face of P then

$$F^\Delta = \{x \in P_r^o : \langle x,y\rangle = r^2 \text{ for all } y \in F\}$$

is a (d-k-1)-face of P_r^o. If $F \subset G$ for two faces, then $F^\Delta \supset G^\Delta$, furthermore $F^{\Delta\Delta} = F$. The similarity $x \mapsto r^2 x$ maps P_1^o onto P_r^o. If v is a vertex of P, then the facet v^Δ of P_r^o lies in a prime L which is perpendicular to the vector v. The distance l of L from 0 is given by $l:r = r:|v|$. Using this the figure 4 is constructed.

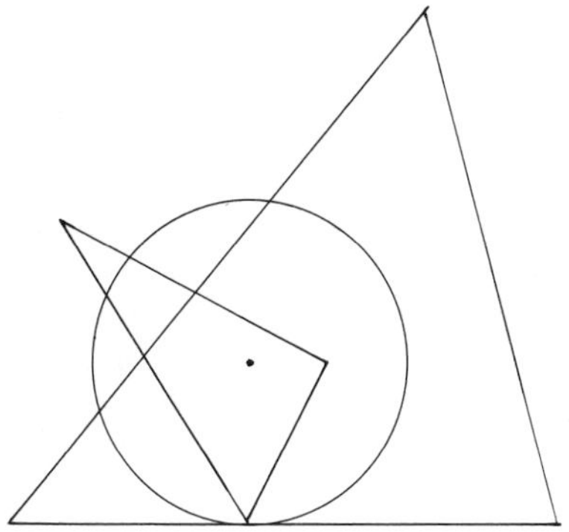

two mutually polar triangles

Figure 4

Proposition: *Let* Q *be a vertex figure of* P *at the vertex* v . *Let the prime* L *through* Q *intersect the line through the vector* v *perpendicularly in the point* c . *The facet* v^Δ *of a polar* P^o *is similar to a polar* Q^c *of* Q *in* L .

Proof: We may assume that $|v|=1$ and that $P^o=P^o_1$. Then

$$v^\Delta = \{x\in \mathbb{R}^d: \langle x,v\rangle=1 \text{ and } \langle x,z\rangle \le 1 \text{ for all } z\in Q\} .$$

This follows from the definitions and the fact that every $y\in P$ can be written as $y=(1-\mu)v+\mu z$ for a suitable $z\in Q$ and $\mu \geq 0$, see figure 5. The equation of L is $\langle x,v\rangle=\lambda$ for some $0<\lambda<1$, see

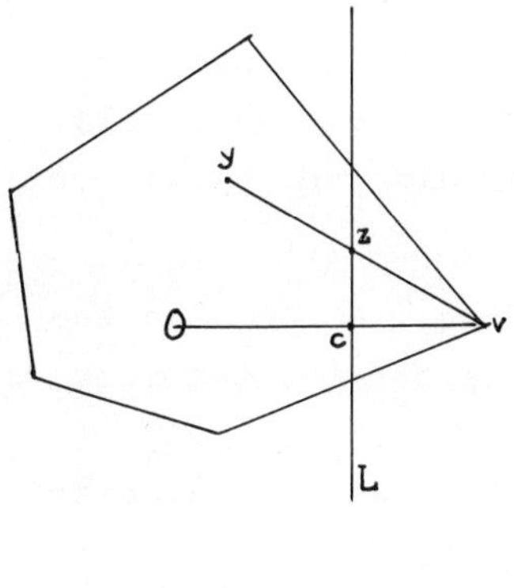

Figure 5

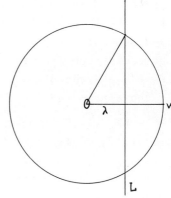

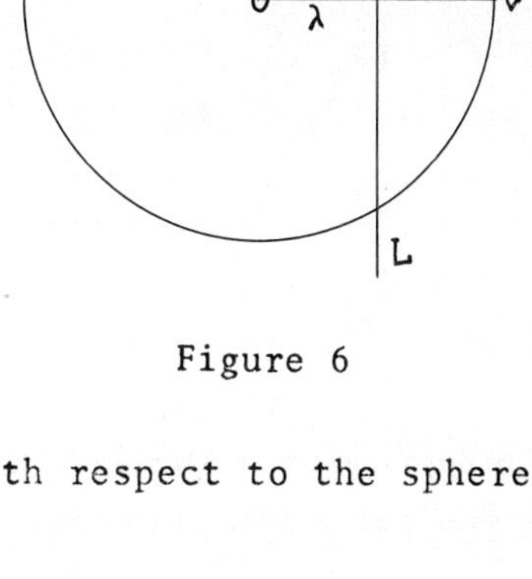

Figure 6

figure 6. The polar of Q in L with respect to the sphere

of radius $\sqrt{1-\lambda^2}$ about c is

$$Q^c = \{x \in \mathbb{R}^d : \langle x,v\rangle = \lambda \text{ and } \langle x,z\rangle \leq 1 \text{ for all } z \in Q\} .$$

The similarity $x \mapsto (x+v)/(1+\lambda)$ maps Q^c onto v^Δ .

§3 Regular Solids

A Euclidean motion which transforms a convex polytope P into itself is called an *automorphism* of P . It permutes the vertices of P and is uniquely determined by this permutation. If $v_1,\ldots,v_q$ are the vertices, the *centre* $c=(v_1+\ldots+v_q)/q$ of P is a fixed point of every automorphism. Every automorphism transforms a [complete] flag of faces into a [complete] flag. The convex polytope P is called *regular* if any two complete flags are transformed into one another by an automorphism of P . Regular convex polytopes will be called *regular solids*.

Each face of a regular solid is itself a regular solid. All faces of the same dimension are congruent to one another. For a fixed dimension k the centres c of all k-faces F have the same distance r_k from the centre 0 of the regular solid P . The vector $\overrightarrow{0c}$ is perpendicular to F . We call $r=r_o$ the *circum radius* of P .

Let $\mathcal{F} = (F_o,\ldots,F_d)$ be a complete flag, let c_k be the centre of F_k . If $c_d=0$ is chosen as origin, the vectors $c_o,\ldots,c_{d-1}$ form a base of $\mathbb{R}^d$. If $\mathcal{F}' = (F'_o,\ldots,F'_d)$ is another complete flag with centres c'_k , an automorphism α of P which transforms $\mathcal{F}$ into $\mathcal{F}'$ satisfies $\alpha(c_k)=c'_k$. Therefore α is uniquely determined by the flags $\mathcal{F}$ and $\mathcal{F}'$. If one complete flag is distinguished, we obtain a 1-1 correspondence between the set of all complete flags of P and the group of all automorphisms of P .

Let v be a vertex of the regular solid P . All edges emanating from v end in vertices lying in one prime L which is perpendicular to the vector v (from the centre O of P to the vertex v). The convex hull of these vertices in L is the *excellent vertex figure* Q of P at v . All excellent vertex figures Q are mutually congruent regular solids of dimension d-1 , if d=dim P .

The following *regularity criterion* is often useful: The convex polytope P is regular if and only if there is a vertex v with the following properties:

(i) There are automorphisms of P which transform v into any other vertex v of P .

(ii) There is a regular vertex figure Q of P at v .

(iii) For every automorphism γ of Q there is an automorphism g of P with $g(v)=v$ and $g|Q=\gamma$.

The only two-dimensional regular solids are the regular p-gons, p=3,4,... . They are denoted by the Schläfli symbol {p} . The *Schläfli symbol* of a three-dimensional regular solid whose two-faces are p-gons and whose vertex figures are q-gons is {p,q} . The Schläfli symbols of higher dimensional regular solids are defined inductively: The (d+1)-dimensional regular solid P has the Schläfli symbol $\{p_1,\ldots,p_d\}$ if its two-faces are p_1-gons and its excellent vertex figures have the Schläfli symbol $\{p_2,\ldots,p_d\}$. Similar regular solids have the same Schläfli symbol.

Let P be a regular solid with edge length 1 and circumradius r . Then $s(P) = 1/2r$ is called the *characteristic ratio* of P ; obviously $0 < s(P) < 1$. Similar solids have the same characteristic ratio. If P is a regular p-gon then $s(P)=\sin \pi/p$. The key for the following results is

Lemma 1: *Let the two-faces of* P *be* p-*gons. Let* Q *be an excellent vertex figure of* P . *Then*

(1) $$s(P)^2 = 1 - \frac{\cos^2\pi/p}{s(Q)^2} \quad .$$

Proof: Let Q be the vertex figure at the vertex V . Let O denote the centre of P and C the centre of Q . Let V' be a vertex of Q . Figure 1 shows that $\sin\varphi = 1/2r$ and $\cos\varphi = r'/l$ hence $s(P)^2 = 1-(r'/l)^2$.

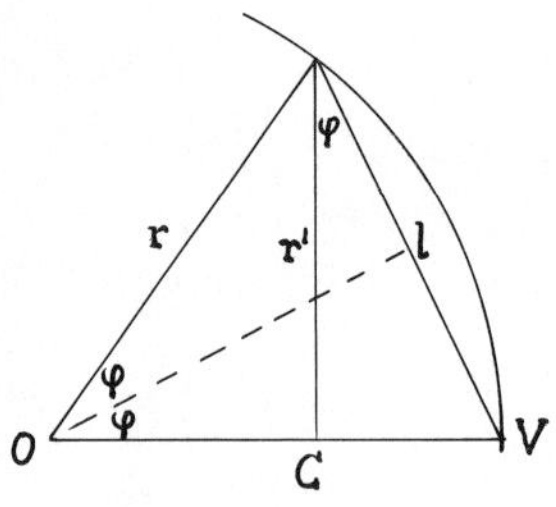

Figure 1

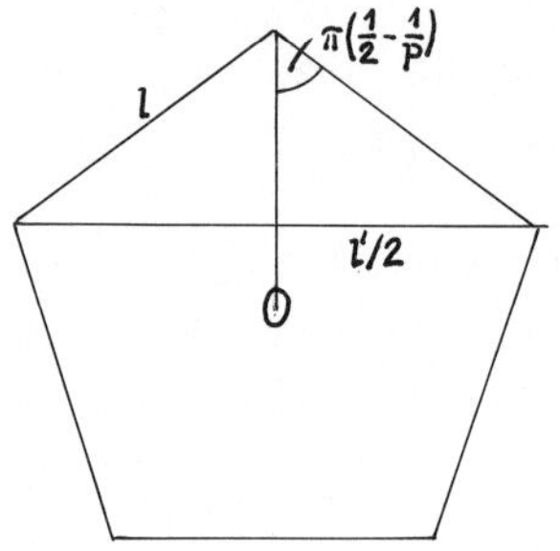

Figure 2

An edge of Q is the vertex figure of a two-face of P containing V , therefore $l' = 2l\cdot\cos\pi/p$ is the edge length of Q , see figure 2 . If this is substituted, the desired formula (1) is obtained because $s(Q) = l'/2r'$.

The lemma implies by induction that the characteristic ratio s(P) depends only on the Schläfli symbol of P . There is a better result:

<u>Theorem 2</u>: *Two regular solids are similar if and only if they have the same Schläfli symbol.*

The proof is by induction on the dimension: Let P and P' be two d-dimensional regular solids in $\mathbb{R}^d$ with the same Schläfli symbol. We may assume that P and P' have the same centre O and at least one vertex V in common. Then they have the same circum-radius r . By induction hypothesis P and P' have similar excellent vertex figures Q and Q' at V . Since $s(P) = \sin\varphi = s(P')$ the vertex figures Q and Q' lie in the same prime L , they have the same centre C and the same circum-radius $r' = r\cdot\sin 2\varphi$, see figure 1 . Therefore Q and Q' are not only similar but congruent:

There is an orthogonal transformation f of $\mathbb{R}^d$ with fixed axis OV such that $f(Q')=Q$. After this transformation P and P' have in common: the centre O, the vertex V, and the excellent vertex figure Q at V. We claim:

(*) P and P' have also in common: all vertices V^*, which are connected to V by an edge, and all the excellent vertex figures Q^* at V^*.

Proof of (*): The vertices V^* are the vertices of the common Q. The reflexion ρ in the prime through O which is perpendicular to VV^* (for a fixed V^*) is an automorphism of P, namely the automorphism which transforms the complete flag $(V,VV^*,F_2,\dots,F_d)$ into the complete flag $(V^*,V^*V, F_2,\dots,F_d)$. The same argument implies that ρ is also an automorphism of P'. Therefore $Q^* = \rho(Q)$ is the common excellent vertex figure of both P and P' at V^*.

For every vertex W of P there is a sequence of vertices $V=V_0,V_1,\dots,V_n=W$ such that V_iV_{i+1} is always an edge of P. The intermediate result (*) implies step by step that $V_0,V_1,\dots,V_n$ are vertices of P', too. Thus P and P' have the same vertices, hence $P=P'$.

Since we consider regular solids up to similarity we can speak of *the* regular solid $\{p_1,\dots,p_{d-1}\}$.

<u>Proposition 3</u>: *The* k-*faces of* $\{p_1,\dots,p_{d-1}\}$ *are* $\{p_1,\dots p_{k-1}\}$

The proof is by induction on the dimension: The excellent vertex figure F' of a k-face F is a (k-1)-face of the vertex figure $\{p_2,\dots,p_{d-1}\}$, therefore $F'=\{p_2,\dots,p_{k-1}\}$ by induction hypothesis. Since the two-faces of F are p_1-gons, $F=\{p_1,p_2,\dots,p_{k-1}\}$.

If P is a regular solid, any polar P^o with respect to a sphere about the centre O of P is called a *dual* of P. All duals of P are mutually similar regular solids. The regular solid P and its dual P^o have the same centre and the same group of automorphisms. A special dual of P is the convex hull of the centres of all facets of P.

Proposition 4: *The dual of* $\{p_1,\ldots,p_{d-1}\}$ *is* $\{p_{d-1},\ldots,p_1\}$.

The proof is by induction on the dimension: Let $Q=\{q_1,\ldots,q_{d-1}\}$ be the dual of $P=\{p_1,\ldots,p_{d-1}\}$. By the proposition at the end of §2 the facets of Q are the duals of the excellent vertex figures of P, therefore by induction hypothesis $\{q_1,\ldots,q_{d-2}\} = \{p_{d-1},\ldots,p_2\}$. Again by the proposition in §2 the dual of the vertex figure $\{q_2,\ldots,q_{d-1}\}$ of Q is a facet of $Q^o=P$, i.e. a $\{p_1,\ldots,p_{d-2}\}$. Therefore by induction hypothesis, the vertex figure $\{q_2,\ldots,q_{d-1}\}$ of Q is $\{p_{d-2},\ldots,p_1\}$.

§4 Enumeration and Realization of Regular Solids

The formula (1) of §3 implies the inequality

(1) $$1/4 \leq \cos^2\pi/p < s(Q)^2$$

for the regular solid $P=\{p,q_2,\ldots,q_{d-1}\}$ with vertex figure $Q=\{q_2,\ldots,q_{d-1}\}$. Here 1/4 comes from $p \geq 3$. We begin with $P=\{p,q\}$. Then $Q=\{q\}$ and $s(Q) = \sin\pi/q$, so that (1) becomes

$$1/4 \leq \cos^2\pi/p < \sin^2\pi/q .$$

This inequality has only five solutions, which are listed in table 1.

P	{3,3}	{3,4}	{4,3}	{3,5}	{5,3}
$s(P)^2$	2/3	1/2	1/3	$(5-\sqrt{5})/10$ ≈ 0.2764	$(3-\sqrt{5})/6$ ≈ 0.1273

Table 1

If $\{p,q,r\}$ is a four-dimensional regular solid, then $\{q,r\}$ occurs in table 1 with a characteristic ratio s such that $1/4 \leq \cos^2\pi/p < s^2$. This restricts the possibilities to the following ones

P	{3,3,3}	{3,3,4}	{4,3,3}	{3,4,3}
$s(P)^2$	5/8	1/2	1/4	1/4

P	{3,3,5}	{5,3,3}
$s(P)^2$	$(3-\sqrt{5})/8 \approx 0.0955$	$(7-3\sqrt{5})/16 \approx 0.0182$

Table 2

In dimensions $d \geq 5$ only three possibilities remain:

P	{3,...,3}	{3,...,3,4}	{4,3,...,3}
$s(P)^2$	$(d+1)/2d$	1/2	$1/(n+1)$

Table 3

Let us first realize the three possibilities of table 3, which occur also in dimension three and four:

Let $e_0=(1,0,..,0)$, $e_1=(0,1,0,..,0)$, ..., $e_d=(0,..,0,1)$ be the unit points of $\mathbb{R}^{d+1}$. Their convex hull T^d is the d-*tetrahedron* with vertices $e_0,\ldots, e_d$. It lies in the prime $x_0+\ldots+x_d = 1$ of $\mathbb{R}^{d+1}$. Special cases are the equilateral triangle T^2 and the regular tetrahedron T^3. Every permutation of the $d+1$ vertices is an automorphism of T^d. Therefore T^d is a regular solid. Its Schläfli symbol is $\{3,\ldots,3\}$.

Let $e_1=(1,0,..,0)$, ..., $e_d=(0,..,0,1)$ be the unit points of $\mathbb{R}^d$. The convex hull of the 2d points $\pm e_1,\ldots,\pm e_d$ is the d-*octahedron* O^d with 2d vertices. Special cases are the square O^2 and the octahedron O^3. If v is a vertex then so is $-v$. A permutation α of the 2d vertices is an automorphism of O^d if and only if $\alpha(-v)=-\alpha(v)$ for every vertex. Therefore O^d is a regular solid. Its Schläfli symbol is $\{3,\ldots,3,4\}$.

The *dual* of the d-tetrahedron is again a d-tetrahedron. The dual of the d-octahedron is the d-*cube* $C^d=\{x\in\mathbb{R}^d: -1\leq x_i\leq+1$ for $i=1,\ldots,d\}$. Special cases are the square C^2 and the cube C^3 . The Schläfli symbol of C^d is $\{4,3,\ldots,3\}$.

It remains to realize the symbol $\{3,5\}$ of table 1 and the symbols $\{3,4,3\}$ and $\{3,3,5\}$ of table 2. The other two $\{5,3\}$ and $\{5,3,3\}$ are duals.

The icosahedron is the regular solid $\{3,5\}$.

Proof: We start at the octahedron with a black and white colouring of its faces, see §1. Every automorphism which leaves this colouring invariant is an automorphism of the icosahedron whose vertices divide the edges of the octahedron according to the golden section. Since these automorphisms acts transitively on the edges of the octahedron they act also transitively on the vertices of the icosahedron. The icosahedron is composed of a pentagonal antiprism with two pentagonal pyramids placed on the two bases of the antiprism. If V is the apex of one of these pyramids the pentagonal base of the pyramid is an excellent vertex figure of the icosahedron at V . Every automorphism of this vertex figure comes from an automorphism of the icosahedron which has V as fixed point. Both descriptions together show that the icosahedron satisfies the regularity criteria (i)-(iii) of §3. Since the faces are triangles and the vertex figures are pentagons, the icosahedron has the Schläfli symbol $\{3,5\}$.

The realization of the four-dimensional solids $\{3,4,3\}$ and $\{3,3,5\}$ is postponed to Chapter II, §3 .

§5 The Rotation Groups of the Platonic Solids

Let the origin of $\mathbb{R}^3$ be the centre of the Platonic solid $P=\{p,q\}$. By definition the rotation group of P consists of the rotations (elements of $SO(3)$) which transform P into itself. It is a subgroup of index 2 of the full automorphism group of P. Since dual solids have the same group, it suffices to study the tetrahedral, octahedral, and icosahedral groups. Unless stated other-wise 'group' will mean 'rotation group'.

Any proper rotation has an axis and an angle. The possibilities are: (1) The axis passes through a vertex and the angle is a multiple of $2\pi/q$. (2) The axis passes through the mid-point of an edge and the angle is π. (3) The axis passes through the centre of a face and the angle is a multiple of $2\pi/p$. These axes are called q-fold, twofold, and p-fold respectively.

With the tetrahedron the axes through vertices and face centres are the same. There are four of them, each being threefold. The mid-edge points occur in antipodal pairs. They yield three twofold axes. We can now list:

The elements of the tetrahedral group:

4 axes through	vertices	yield 2 elements each (of order 3)	8
3 axes through	mid-edge points	yield 1 element each (of order 2)	3
		the identity	1
		order of the tetrahedral group	12

With the octahedron and the icosahedron the vertices, mid-edge points and face centres occur in antipodal pairs. Hence there are $f/2$ many p-fold axes passing through the face centres, $v/2$ many q-fold axes passing through the vertices, and $e/2$ many twofold axes passing through the mid-edge

points. Here v,e,f denote the numbers of vertices, edges, and faces respectively, see the table in §2.

We obtain:

The elements of the octahedral group:

3 axes through	vertices	yield 3 elements each (2 of order 4, 1 of order 2)	9
4 axes through	face centres	yield 2 elements each (of order 3)	8
6 axes through	mid-edge points	yield 1 element each (of order 2)	6
		the identity	1
		order of the octahedral group	24

The elements of the icosahedral group:

6 axes through	vertices	yield 4 elements each (of order 5)	24
10 axes through	face centres	yield 2 elements each (of order 3)	20
15 axes through	mid-edge points	yield 1 element each (of order 2)	15
		the identity	1
		order of the icosahedral group	60

We shall identify the tetrahedral, octahedral and icosahedral groups with *permutation groups*. The full automorphism group T^* of the tetrahedron in the symmetric group $\mathcal{S}_4$ of all permutations of the four vertices. The even permutations are rotations. Therefore *the tetrahedral (rotation) group is the alternating group* $\mathcal{A}_4$.

With the *octahedron* the eight faces occur in four pairs of parallel faces. The elements of the octahedral group permute these four pairs effectively, i.e. no element $\neq$ id yields the identical permutation. Since there are 24 possible permutations, all occur. Thus *the octahedral group is the symmetric group* $\mathcal{S}_4$.

The edges of the *icosahedron* occur in 5 classes of 6 edges each: Two edges belong to the same class if and only if they are parallel or perpendicular to one another. Five edges with a common vertex belong to five different classes. Similarly, five edges which belong to two faces with a common edge, belong to five different classes.

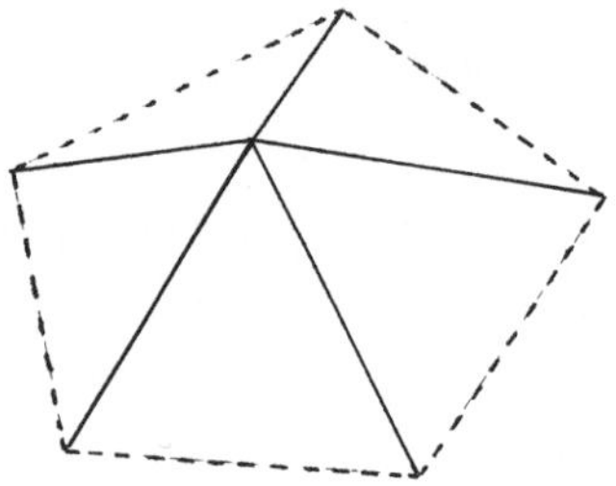

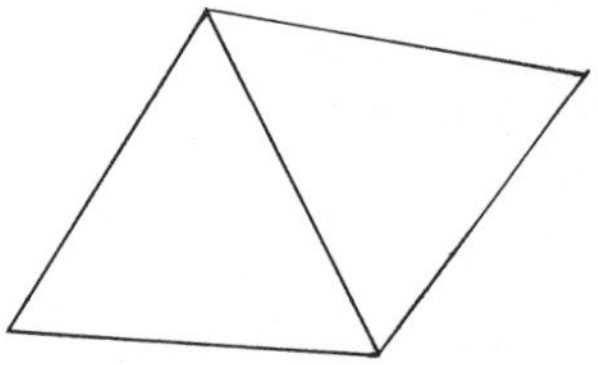

The elements of the icosahedral group permute the five classes. The rotations about an axes through a vertex yield cyclic permutations of order five as we see from the representing five edges at this vertex. The rotation about an axis through a mid-edge point yields two transpositions of the five classes as we see from the second figure above. Finally a rotation about an axis through a face centre yields a cyclic permutation of the three classes represented by the sides of the (triangular) face. The other two classes cannot be permuted because the rotation has order three. This inspection shows: The icosahedral group may be considered as a subgroup of $\mathcal{S}_5$, only even permutations occur. Since there are 60 even permutations, all of them occur, and we have proved: *The icosahedral group is the alternating group* $\mathcal{A}_5$.

§6 Finite Subgroups of the Rotation Group SO(3)

Besides the groups of the regular solids there are two more infinite series of finite subgroups of SO(3) : There are *cyclic* subgroups of any order $p < \infty$. Each of them consists of all rotations about a fixed axis with angles $2\pi j/p$, $j=1,\dots p$. Any two such groups are conjugate in SO(3) .

Given three mutually perpendicular lines through the origin there is the finite subgroup which leaves each line invariant. It consists of the 180°-rotations about the three lines and the identity. This group of order 4 is called the 2-*dihedral* group or the four-group. Any two of them are conjugate in SO(3) .

For $p \geq 3$ consider a plane regular p-gon with center at the origin. The p-dihedral group consists of all rotations which leave this p-gon invariant. With this group the line through the origin perpendicular to the p-gon is a p-fold axis. The other axes of the group are twofold. They pass through the vertices and mid-edge points of the p-gon and there are p of them. Hence the p-*dihedral* group is of order 2p . Again any two such groups are conjugate in SO(3) .

Remark: The plane p-gon may be considered as a degenerate regular solid with its face counted twice. Therefore it is also called a *di*hedron.

Up to conjugacy the finite subgroups of SO(3) *are the following ones: The cyclic groups of order* $p=2,3,\dots,$ *the dihedral groups of order* 2p $(p=2,3,\dots.)$, *the tetrahedral, octahedral, and icosahedral groups.*

Proof: Since SO(3) transforms the unit sphere S^2 into itself we restrict attention to the points $P \in S^2$. Let $G < SO(3)$ be a finite subgroup. For each $P \in S^2$ we have the *isotropy subgroup*

$$G_P = \{g \in G : gP = P\} < G$$

and the *orbit*

$$G\cdot P = \{gP : g \in G\} \subset S^2 .$$

The sphere S^2 is the disjoint union of the orbits. The order $n(P)$ of G_P divides the order N of G. The quotient $c = N/n(P)$ is the cardinality of $G\cdot P$ because

$$G/G_P \to G\cdot P , \ g\cdot G_P \mapsto gP ,$$

is a well defined bijection. This implies that $n(P) = n(Q)$ if P and Q lie in the same orbit. It $n(P) > 1$ the orbit $G\cdot P$ is called *exceptional*. We consider the set

$$\mathfrak{M} = \{ (g,P) : g \in G\setminus\{1\} , P \in S^2 , gP = P\} .$$

If we fix $g \in G\setminus\{1\}$ there are exactly two points P and Q with (g,P) and (g,Q) lying in $\mathfrak{M}$, namely the antipodal points where the axis of g intersects the sphere S^2. This implies

(1) $$\#\mathfrak{M} = 2(N-1) .$$

On the other hand, if we fix $P \in S^2$, then $(g,P) \in \mathfrak{M}$ if and only if $g \in G_P\setminus\{1\}$. Thus every $P \in S^2$ contributes $n(P)-1$ elements of $\mathfrak{M}$ and we need only consider the points P which lie in exceptional orbits. Since $\#\mathfrak{M} < \infty$, only finitely many exceptional orbits $\Sigma_1,\ldots,\Sigma_k$ can occur. Let $c_i = \#\Sigma_i$ and let $n_i = n(P)$ for some $P \in \Sigma_i$. The elements of Σ_i contribute $c_i(n_i-1)$ elements of $\mathfrak{M}$ and hence

(2) $$\#\mathfrak{M} = \sum_{i=1}^{k} c_i(n_i-1) .$$

From (1) and (2) we obtain $2(N-1) = \sum_{i=1}^{k} c_i(n_i-1)$. This restricts the possibilities: Division by $N = c_i n_i$ yields

(3) $$\sum_{i=1}^{k} \frac{1}{n_i} = \frac{2}{N} + k-2 .$$

Since $n_i \geq 2$ the left hand side is $\leq \frac{1}{2}k$ and therefore $k < 4$, actually $k = 2$ or 3 because $k = 1$ is also impossible with (3). For $k=2$ we obtain $1/n_1 + 1/n_2 = 2/N$, hence $n_1 = n_2 = N$ i.e. $G = G_P$ for some P. Thus G is cyclic.

For $k=3$ we have

$$\frac{1}{n_1} + \frac{1}{n_2} + \frac{1}{n_3} = \frac{2}{N} + 1 \text{ , in particular } \frac{1}{n_1} + \frac{1}{n_2} + \frac{1}{n_3} > 1 \text{ .}$$

For $n_1 \leq n_2 \leq n_3$ the inequality has only the following solutions: $(n_1,n_2,n_3) = (2,2,q)$ with $q \geq 2$, $=(2,3,3)$, $= (2,3,4)$, $= (2,3,5)$. This yields the following table for the possible n_i, c_i, and N:

n_1	2	2	2	2
n_2	2	3	3	3
n_3	$q \geq 2$	3	4	5
c_1	q	6	12	30
c_2	q	4	8	20
c_3	2	4	6	12
N	2q	12	24	60
type	D_q	T	O	I

The dihedral, tetrahedral, octahedral, and icosahedral groups belong to the types D_q, T, O, and I respectively. It remains to show, that *any* finite subgroup of type D_q is a q-dihedral group and similarly with the other types.

Let G be of type D_2. Then the exceptional orbits consist of two points each, $\Sigma_i = \{P_i, Q_i\}$, $i=1,2,3$. We claim $Q_i = -P_i$ and $\vec{P}_i \perp \vec{P}_j$ (perpendicular) for $i \neq j$. Indeed, there is exactly one $g \in G \setminus \{1\}$ with $gQ_i = Q_i$. Then $gP_i = P_i$ because any $g \in G$ leaves Σ_i invariant. Since the fix-point set of any $g \in SO(3)-\{1\}$ consists of exactly two antipodal points, $Q_i = -P_i$. Furthermore, $gP_j = -P_j$ for $j \neq i$. Since g leaves the inner product invariant, we obtain $\langle \vec{P}_i, \vec{P}_j \rangle = \langle \vec{P}_i, -\vec{P}_j \rangle$. Thus the three axis through the origin and P_1, P_2, P_3 are mutually perpendicular. Each of them is left invariant by G. Therefore G is a subgroup of the corresponding 2-dihedral group. Since both groups have the same order 4, they coincide.

For any group G of type D_q $(q \geq 3)$, T, O, or I we shall find a q-dihedron, a tetrahedron, an octahedron, or an icosahedron respectively, which remains invariant under G. Then G will coincide with the corresponding dihedral, tetrahedral, octahedral, or icosahedral group.

If G has type D_q $(q \geq 3)$ we choose a point $P \in \Sigma_3$. As with D_2 we may show that $\Sigma_3 = \{P, -P\}$ and $\vec{P} \perp \vec{Q}$ for any $Q \in \Sigma_1 \cup \Sigma_2$. Therefore the convex hull of Σ_1 is a q-gon lying in the plane through the origin perpendicular to $\vec{P}$. Since the cyclic group G_P of order q permutes the points of Σ_1 in cyclic order, the q-gon is regular. It is the q-dihedron we were looking for.

If G has type T we consider the four points of Σ_3. For any $P \in \Sigma_3$ the isotropy group G_P has order 3. Hence it permutes the three points of $\Sigma_3 \setminus \{P\}$ in cyclic order. Thus they are the vertices of an equilateral triangle. Since P was arbitrary, any three of the four points of Σ_3 form an equilateral triangle. Hence the convex hull of Σ_3 is a tetrahedron which is invariant under G.

If G has type O we consider the six points of Σ_3 and claim: $\Sigma_3 = \{P,Q,R,-P,-Q,-R\}$ with $\vec{P}, \vec{Q}$, and $\vec{R}$ mutually perpendicular. Then the convex hull of Σ_3 will be an octahedron, which is invariant under G. In order to prove our claim we observe that for any $P \in \Sigma_3$ the isotropy group G_P, which is cyclic of order four, permutes the remaining five points of Σ_3. Then one of them must stay fixed, i.e. it must be $-P$. So we know already that the points of Σ_3 occur in antipodal pairs. Given now any two points P and $Q \neq -P$ from Σ_3 we have just observed that G_P permutes $\{Q,R,-Q,-R\}$ in cyclic order. In particular, there is a $g \in G_P$ with $gQ = -Q$. Since $\langle \vec{P}, \vec{Q} \rangle = \langle g\vec{P}, g\vec{Q} \rangle = \langle \vec{P}, -\vec{Q} \rangle$, it follows $\langle \vec{P}, \vec{Q} \rangle = 0$.

Finally let G be of type I . Then the orbit Σ_3 consists of 12 points. For any $P \in \Sigma_3$ the isotropy group G_P , which is cyclic of order 5 , permutes the remaining 11 points. Then one of them must be a fixed point, i.e. it must be $-P$. Thus the points of Σ_3 occur in antipodal pairs. We consider now the action of G_P on $\Sigma_3 \setminus \{P,-P\}$. There are two orbits of five points each. The convex hull of each orbit is a regular pentagon Π resp. Π' . Among the vertices of a pentagon no antipodal pairs can occur. Therefore $\Pi' = -\Pi$ is the antipodal one of Π . Let Π be the pentagon which is closer to P than Π' . The convex hull of $\{P\} \cup \Pi$ is a right pyramid. Its base Π is placed on one base of the antiprism, which is the convex hull of $\Pi \cup (-\Pi)$. On the other base the antipodal pyramid, i.e. the convex hull of $\{-P\} \cup (-\Pi)$, is placed. As we compare this situation with the construction of the icosahedron (§1) we see that the convex hull of Σ_3 is an icosahedron. It remains to prove that all its edges have equal length: So far three lengths may occur: The lateral edges of the two pyramids, the lateral edges of the antiprism and the sides of the two pentagons. Since P is the top of a pyramid, all edges from P have equal length. At a vertex Q of Π edges of all three kinds come together. But we started with an arbitrary $P \in \Sigma_3$. Therefore, as from P , all edges from Q have equal length.

For later reference the results of this section are summarized in the following tables. They list the exceptional orbits, i.e. the two fixed points $\pm P$ for the cyclic group and

E = {mid-edge points} , V = {vertices} , F = {face centers}

for the other groups, the number c of the elements of the orbit, the order n of the corresponding cyclic isotropy subgroup, and the number a of the corresponding axes.

cyclic order N	P	-P
c	1	1
n	N	N
a	1	

dihedral order 2q	E	V	F	
c	q	q	2	
n	2	2	q	
a	q/2	q/2	1	q even
	q		1	q odd

tetrahedral order 12	E	V	F
c	6	4	4
n	2	3	3
a	3	4	

octahedral order 24	E	V	F
c	12	6	8
n	2	4	3
a	6	3	4

icosahedral order 60	E	V	F
c	30	12	20
n	2	5	3
a	15	6	10

§7 The Normal Subgroups

It is well known andeasily checked that the cyclic group of order N has exactly one (normal) subgroup of index r for every r which divides N . Here are the corresponding results for the other finite subgroups of SO(3) .

q-*Dihedral Group:*

(1) Every subgroup consisting of rotations about the face centre axis only is normal.

If q is odd, there are no other proper normal subgroups. If q is even, there are two more possibilities:

(2) The rotations about the face centre axis, whose angles are multiples of $4\pi/q$ together with the rotations about the vertex axes form a normal subgroup.

(3) As in (2) with vertex axes replaced by mid-edge point axes.

Tetrahedral Group:

There is only one proper normal subgroup consisting of the identity and the three π-rotations about the mid-edge point axes. It has index 3 .

Octahedral Group:

There are two proper normal subgroups. The smaller one consists of the identity and the π-rotations about the vertex axes. The corresponding factor group is the non-abelian group of order 6 $(\cong \mathcal{S}_3)$. The bigger normal subgroup has index 2 and contains in addition the 8 rotations about the 4 face centre axes.

Icosahedral Group:

There are no proper normal subgroups.

The proofs of these results rely on the following fact: Let $H < \Gamma$ be a normal subgroup, let $h \in H$, $h \neq id$. Then $\gamma h \gamma^{-1} \in H$ for every $\gamma \in \Gamma$. If s denotes the axis of h , then $\gamma(s)$ is the axis of $\gamma h \gamma^{-1}$. The angles of h and $\gamma h \gamma^{-1}$ are the same. This implies: If H contains a φ-rotations about s then H contains all $m\varphi$-rotations about all axes which have the same type (face centres etc.) as s for all $m \in \mathbb{Z}$.

We shall consider the icosahedral group Γ in detail; the reader can prove the results for the other groups similarly. The table at the end of §6 is used. A normal subgroup H of

Γ contains either all 15 rotations $\neq$ id about the mid-edge point axes or none of them; it contains either all 24 rotations $\neq$ id about the vertex axes or none of them, and it contains either all 20 rotations $\neq$ id about the face centre axes or none of them, finally $id \in H$; thus $\# H = 15\alpha+24\beta+20\gamma+1$ with $\alpha,\beta,\gamma \in \{0,1\}$. But this $\# H$ never divides $60 = \# \Gamma$ properly.

§8 Generators and Relations for the Finite Subgroups of SO(3)

The cyclic group of order n is obviously generated by one $2\pi/n$-rotation β with $\beta^n=1$. The other finite subgroups of SO(3) are generated by two elements β,γ each, which are chosen as follows:
With the q-dihedral group γ is the $2\pi/q$-rotation about the face centre axis and β is the π-rotation about some other axis, see figure 1 for $q=3$ and $=4$. With the groups of the Platonic solids

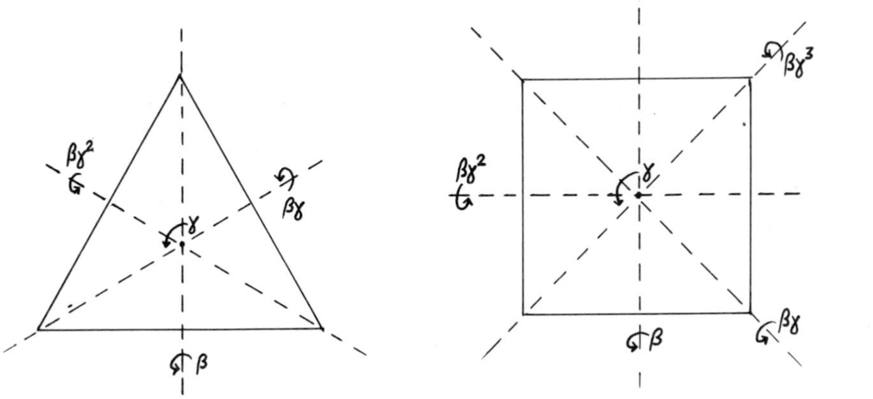

Figure 1

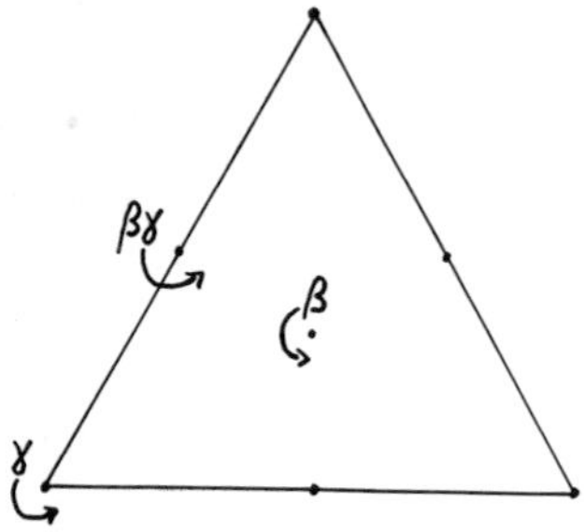

Figure 2

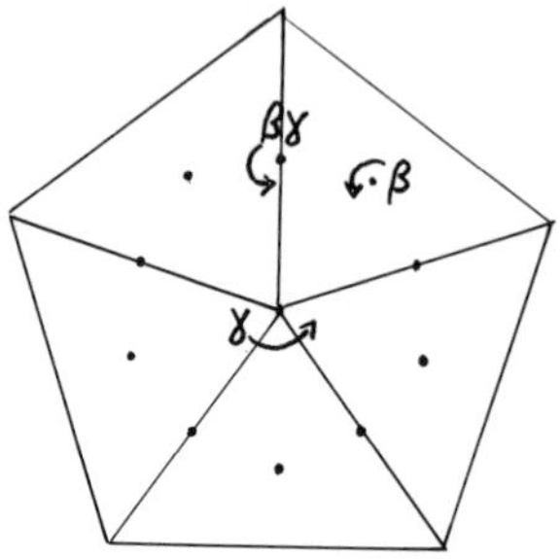

Figure 3

β is the $2\pi/3$ rotation about some face centre axis and γ is the $2\pi/r$- rotation about the axis through one vertex of the same face $(r=3,4,5)$, see figure 2. *The finite subgroups of* SO(3) *are presented by these generators and the following relations:*

type of $\Gamma<SO(3)$	generators	relations
cyclic of order n	β	$\beta^n=1$
q-dihedral	β,γ	$\beta^2=\gamma^q=(\beta\gamma)^2=1$
tetrahedral	β,γ	$\beta^3=\gamma^3=(\beta\gamma)^2=1$
octahedral	β,γ	$\beta^3=\gamma^4=(\beta\gamma)^2=1$
icosahedral	β,γ	$\beta^3=\gamma^5=(\beta\gamma)^2=1$

The proof consists of three parts: (a) The relations hold true. (b) The generators β,γ suffice. (c) The relations suffice.

Part (a) is the easiest: The product $\beta\gamma$ is a π-rotation as indicated in figures 1 and 2. Hence $(\beta\gamma)^2=1$. The other relations are obvious.

Part (b): With the dihedral group the powers β^m yield all rotations about the face centre axes. The π-rotations about the other axes have the form $\beta\gamma^m$, see figure 1. For the groups of the Platonic solids figures 2 and 3 are considered. Here figure 3 shows part of the surface of the icosahedron. All rotations about the axes through the marked mid-edge points are obtained from $\beta\gamma$ by conjugation with the powers of β (figure 2) and γ (figure 3). Similarly all rotations about the axes through the marked vertices of figure 2 are obtained from γ^m by conjugation with the powers of β, and all rotations about the axes through the marked face centres of figure 3 are obtained from β and β^2 by conjugation with the powers of γ. In this way all elements of the group are obtained from β and γ.

Part (c) is the most difficult one: We consider only the icosahedral group, the other cases are similar. The figure 4 shows part of the surface of the icosahedron barycentrically subdivided.

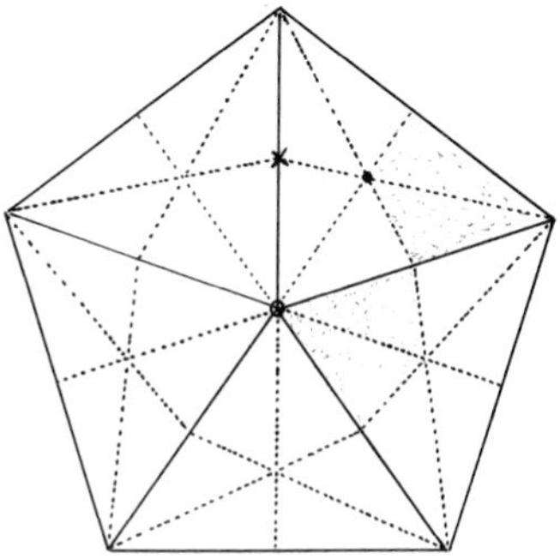

Figure 4

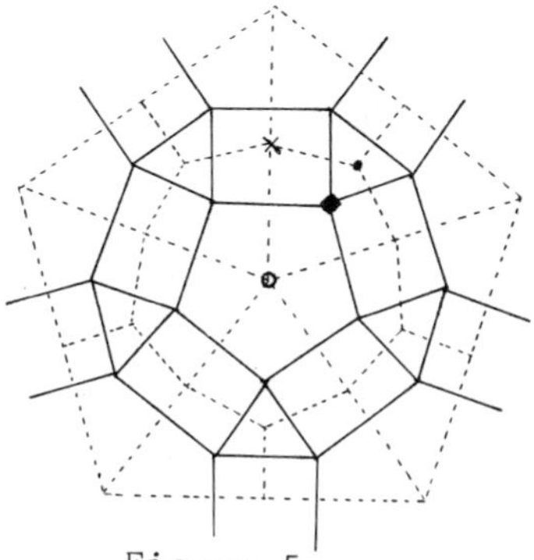

Figure 5

β rotates about •, γ rotates about ∘, $\beta\gamma$ rotates about ×. The base point P used later is at ■.

Every two adjacent triangles of the subdivision form a "kite". Thus a cell decomposition of the sphere consisting of 60 kites is obtained. Each kite is a fundamental region of the Γ-action, i.e. every orbit passes through the kite and any two interior points of the kite belong to different orbits. We pass to the dual cell decomposition, see figure 5. Its vertices are the 60 barycenters of the kites. There are 120 edges. The 2-cells are 12 pentagons, 30 quadrangles and 20 triangles.

A finite sequence of vertices $P_1 \ldots P_n$ is called an edge path if P_i and P_{i+1} are connected by an (unique) edge of the subdivision. The edge path shall be viewed as sequence of its oriented edges, see figure 6.

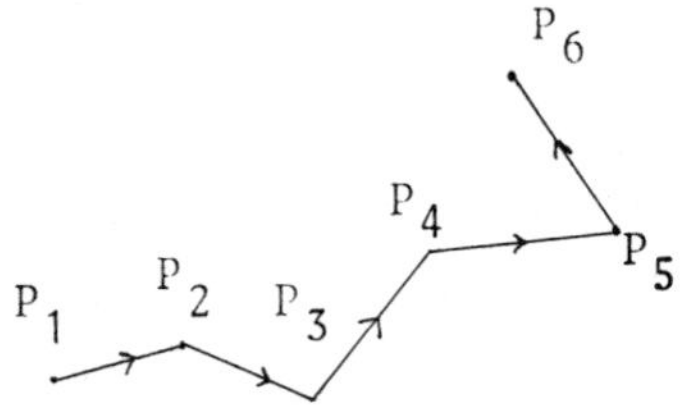

Figure 6

We consider the following operations transforming edge paths:

(1) $\ldots PQP \ldots \leftrightarrow \ldots P \ldots$, see figure 7

(2) $\ldots PQ_1 \ldots Q_r \ldots \leftrightarrow \ldots P \ldots$ if $Q_1, \ldots, Q_r$ are the vertices of a cell in cyclic order $(r=3,4,5)$ and $P=Q_r$, see figure 8

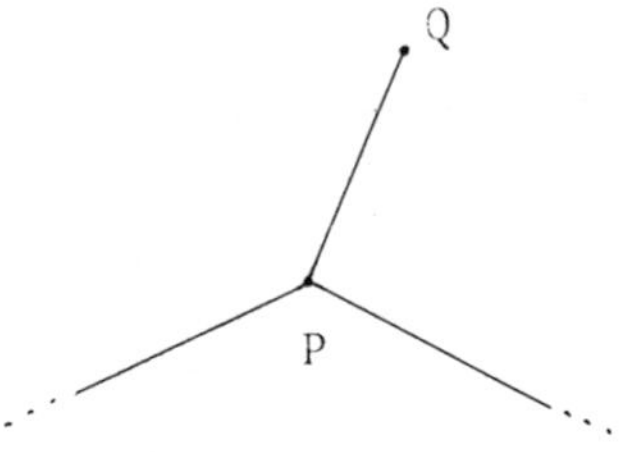

Figure 7

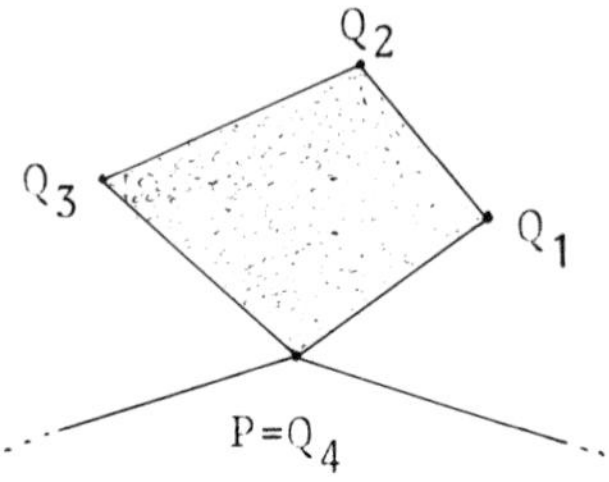

Figure 8

An edge path of the cell decomposition of a surface can be considered as continuous path. Two results from combinatorial topology will be used. The first one says: Two edge paths of the same cell decomposition are homotopic (in the sense of continuous topology) if and only if they can be transformed into one another by finitely many operations (1) and (2). The second result says: The surface of a convex polytope is simply connected, i.e. every continuous path from and to the same point P (a 'loop') is homotopic to the constant path. For both results, see e.g. Seifert-Threlfall, in particular §§45,46. Both results imply:

Every edge path $P_1 \ldots P_n$ with $P_1 = P_n$ can be transformed into the constant path P_1 using (1) and (2) finitely many times.

Let $\alpha_1 \ldots \alpha_n$ be a word consisting of $\alpha_i \in \{\beta, \gamma, \beta^{-1}, \gamma^{-1}\}$. It represents an element in the group Γ. The empty word represents 1. The word $w = \alpha_1 \ldots \alpha_n$ determines the edge path $[\alpha_1 \ldots \alpha_n] =: P\ \alpha_1(P)\ \alpha_1\alpha_2(P)\ \ldots\ \alpha_1 \ldots \alpha_n(P)$. Here P is the base point as in figure 5. Actually every edge path starting at P has this form. If $\alpha_1 \ldots \alpha_n = 1$ in Γ the path is a loop. Using the result stated above the loop can be transformed into the constant loop $[\emptyset] = P$ using (1) and (2). The corresponding operations for words are

(1') $\ldots \alpha_i \alpha \alpha^{-1} \alpha_{i+1} \ldots \leftrightarrow \ldots \alpha_i \alpha_{i+1} \ldots$
for some $\alpha \in \{\beta, \gamma, \beta^{-1}, \gamma^{-1}\}$

(2') $\ldots \alpha_i \beta^{\pm 3} \alpha_{i+1} \ldots \leftrightarrow \ldots \alpha_i \gamma^{\pm 5} \alpha_{i+1} \ldots \leftrightarrow$
$\alpha_i (\beta\gamma)^{\pm 2} \alpha_{i+1} \ldots \leftrightarrow \ldots \alpha_i \alpha_{i+1} \ldots$

depending on the cell (triangle, pentagon, quadrangle) used in (2). Thus a word $\alpha_1 \ldots \alpha_n$ which represents $\alpha_1 \ldots \alpha_n = 1$ in Γ can be transformed into the empty word using the relations given in the table.

CHAPTER II

FINITE SUBGROUPS OF $SL(2,\mathbb{C})$ AND INVARIANT POLYNOMIALS

In the first chapter the finite subgroups Γ of $SO(3)$ have been classified using their action on the Euclidean two-sphere S^2 . Now S^2 is considered as Riemannian sphere, i.e. as Gauss plane $\mathbb{C}$ of the complex variable z compactified by one point ∞ . Then $SO(3)$ and hence every finite subgroup Γ acts by projective (linear, homographic) transformations

$$z \mapsto \frac{az+b}{cz+d} \quad \text{with} \quad ad-bc \neq 0 .$$

We may assume $ad-bc=1$. Thus in §1 a close relation between the finite subgroups Γ of $SO(3)$ and the finite subgroups G of the group $SL(2,\mathbb{C})$ of all complex 2×2 matrices $\begin{pmatrix} a & b \\ c & d \end{pmatrix}$ with determinant 1 is established: To every finite $\Gamma < SO(3)$ there corresponds a 'binary' $G < SL(2,\mathbb{C})$ with twice as many elements.

A 'base' vector $0 \neq v \in \mathbb{C}^2$ determines the embedding $G \hookrightarrow \mathbb{C}^2$, $g \mapsto g\cdot v$. In §3 the convex hull $\operatorname{ch} G$ in $\mathbb{C}^2 = \mathbb{R}^4$ is studied. In three cases $\operatorname{ch} G$ proves to be a regular 4-dimensional solid, namely $\{3,3,4\}$, $\{3,4,3\}$, and $\{3,3,5\}$. Together with the dual $\{5,3,3\}$ this completes the realization of all possible regular solids, which we began in I§4.

For a suitably chosen Hermitean metric on $\mathbb{C}^2$ the action of G leaves the unit sphere $S^3 \subset \mathbb{C}^2$ invariant. In §4 the orbit spaces S^3/G , which are compact 3-manifolds, are investigated from the viewpoint of algebraic and combinatorial topology.

From §6 on, the action of G on $\mathbb{C}^2$ is studied using the results of the first chapter, in particular the geometry of the Platonic solids. This leads to a description of the $\mathbb{C}$-algebra S^G of G-invariant polynomials in two variables by three generators x,y,z and one relation $\varphi(x,y,z)=0$ in §8, table 4.

Finally in §9 a general result about S^G for arbitrary finite subgroups $G < GL(n,\mathbb{C})$ is proved and used in order to describe the orbit space $\mathbb{C}^n/G$ as affine variety $V \subset \mathbb{C}^r$. In our case $G < SL(2,\mathbb{C})$ this is the complex surface $V = \{(x,y,z) \in \mathbb{C}^3 : \varphi(x,y,z)=0\}$.

The transition from the finite subgroups of $SO(3)$ to the finite subgroups G of $SL(2,\mathbb{C})$ as well as the description of the algebra S^G is due to Chapter II of Klein's Lectures on the Icosahedron. Our semi-invariants are his invariants. Klein obtains all (semi-)invariants by explicit calculations in suitably chosen coordinates using the general theory of covariants. We prefer to deduce the same results as far as possible from the geometry of the Platonic soldis without any coordinate calculations. Other presentations of Klein's results have been given by DuVal and Springer. Coxeter (1) realizes the regular 4-dimensional solids without using the finite subgroups of $SL(2,\mathbb{C})$. Later DuVal and Coxeter (2) use these groups. The topological investigation of the 3-manifolds S^3/G goes back to Seifert and Threlfall (1).

§1 Finite Subgroups of $SL(2,\mathbb{C})$

Let $S^2 \subset \mathbb{R}^3$ denote the unit sphere. It will be considered as Riemannian sphere in the sense of complex analysis. Every element $\alpha \in SO(3)$ is an orientation preserving isometry of S^2, in particular an orientation preserving conformal mapping and therefore a biholomorphic mapping.

In order to introduce complex coordinates the Riemannian sphere will be described as *complex projective line* $\mathbb{P}$, i.e. as

$\mathbb{C}^2 \setminus \{0\}$ modulo the equivalence relation

$$z \sim w \text{ if and only if } w = \lambda z \text{ for some } \lambda \in \mathbb{C}^\times .$$

The equivalence class of $z=(z_0,z_1) \in \mathbb{C}^2 \setminus \{0\}$ will be denoted by $[z] = [z_0:z_1] \in \mathbb{P}$. The action of the general linear group $GL(2,\mathbb{C})$ on $\mathbb{C}^2 \setminus \{0\}$ induces an action on $\mathbb{P}$ defined by

$$g\cdot[z]=[g\cdot z] \quad \text{for } g \in GL(2,\mathbb{C}) \text{ and } z \in \mathbb{C}^2 \setminus \{0\} .$$

The normal subgroup

$$Z = \{\lambda E: \lambda \in \mathbb{C}^\times\} \quad \text{with} \quad E = \begin{pmatrix} 1 & 0 \\ 0 & 1 \end{pmatrix}$$

acts trivially on $\mathbb{P}$. The quotient group

$$PGL(2,\mathbb{C}) = GL(2,\mathbb{C})/Z$$

acts effectively. It is called the *projective linear group*. As far as the action on $\mathbb{P}$ is concerned it suffices to consider the *special linear group* $SL(2,\mathbb{C}) < GL(2,\mathbb{C})$ of matrices A with $\det A = 1$ because every element of $PGL(2,\mathbb{C})$ is represented by some $A \in SL(2,\mathbb{C})$. The kernel of the canonical epimorphism $\rho: SL(2,\mathbb{C}) \to PGL(2,\mathbb{C})$ consists of E and $-E$. From complex analysis we have the result:

Every biholomorphic mapping $\alpha: \mathbb{P} \to \mathbb{P}$ *belongs to* $PGL(2,\mathbb{C})$.

In particular, $SO(3)$ is a subgroup of $PGL(2,\mathbb{C})$. The inverse image

$$SU(2) = \rho^{-1}(SO(3)) < SL(2,\mathbb{C})$$

is called the *special unitary group*. This is not the usual definition. But together with the following proposition 1 it will suffice for our purposes. The following §2 presents another approach, using quaternions, including a proof of

Proposition 1: *Every finite subgroup of* $SL(2,\mathbb{C})$ *is conjugate to a subgroup of* $SU(2)$.

Hence in order to classify the finite subgroups of $SL(2,\mathbb{C})$ up to conjugation we may restrict to finite subgroups of $SU(2)$. The following simple lemma will be useful.

Lemma 2: *If* $A \in SL(2,\mathbb{C})$ *has order* 2 *then* $A=-E$.

Proof: Let $A = \begin{pmatrix} a & b \\ c & d \end{pmatrix}$. Then $\begin{pmatrix} d & -b \\ -c & a \end{pmatrix} = A^{-1} = A = \begin{pmatrix} a & b \\ c & d \end{pmatrix}$. This implies $a=d=\pm 1$ and $b=c=0$.

If $\Gamma < SO(3)$ is a finite subgroup $\rho^{-1}(\Gamma) < SU(2)$ and any $G < SL(2,\mathbb{C})$ which is conjugate to $\rho^{-1}(\Gamma)$ is called a *binary* subgroup corresponding to Γ . This G has twice as many elements as Γ . A finite subgroup of $SL(2,\mathbb{C})$ is binary if and only if it contains $-E$.

<u>Lemma 3</u>: *A finite subgroup* $G < SL(2,C)$ *is not binary if and only if it is cyclic of odd order.*

Proof: We may assume that $G < SU(2)$ by proposition 1 . Since $-E \notin G$, the projection ρ maps G isomorphically onto the finite subgroup $\rho(G) < SO(3)$. Lemma 2 implies that G and therefore $\rho(G)$ contain no elements of order 2 . It follows from the classification of finite subgroups of $SO(3)$, see I§6, that $\rho(G)$ and thus G is cyclic of odd order.

Using the classification of finite subgroups of $SO(3)$ again, we obtain

<u>Theorem</u>: *There are the following conjugacy classes of finite subgroups of* $SL(2,\mathbb{C})$: *The cyclic groups of order* $1,2,\ldots$, *the binary dihedral subgroups of order* $4q$ $(q=2,3,\ldots)$, *the binary tetrahedral, octahedral and icosahedral groups of orders* 24, 48 *and* 120 *respectively.*

§2 Quaternions and Rotations

The usual definition of $SU(n)$ is the following one: For a matrix $A=(a_{ij})$ let $A^*=(\bar{a}_{ji})$ denote the transposed conjugate matrix. Then $U(n) = \{A \in GL(n,\mathbb{C}): A^*=A^{-1}\}$ and $SU(n) = U(n) \cap SL(n,\mathbb{C})$. If $\langle -,- \rangle$ denote the canonical Hermitean inner product, $U(n)$ consists of those A which preserve $\langle -,- \rangle$, i.e. $\langle Az,Aw \rangle = \langle z,w \rangle$. In order to prove proposition 1 of the preceding section two facts are used:

(a) For any finite subgroup G of $GL(n,\mathbb{C})$ there is a G-invariant Hermitean inner product $\langle -,- \rangle_G$ on $\mathbb{C}^n$. It can be obtained from the canonical inner product $\langle -,- \rangle$ by averaging:

$$\langle x,y \rangle_G = \frac{1}{\text{ord } G} \sum_{g \in G} \langle gx,gy \rangle .$$

(b) Given any two Hermitean inner products $\langle -,- \rangle$ and $\langle -,- \rangle_*$ on $\mathbb{C}^n$, there is a $T \in GL(n,C)$, such that

(1) $\qquad \langle x,y \rangle_* = \langle Tx,Ty \rangle$ for $x,y \in \mathbb{C}^n$.

For (b) consider two orthonormal bases $\mathcal{B}$ with respect to $\langle -,- \rangle$, and $\mathcal{B}_*$ with respect to $\langle -,- \rangle_*$. There is a $T \in GL(n,\mathbb{C})$ such that $T\mathcal{B}_*=\mathcal{B}$. Then (1) holds true for any elements x and y of $\mathcal{B}$ and hence for arbitrary x,y in $\mathbb{C}^n$.

Given a finite subgroup G of $GL(n,\mathbb{C})$ we choose a G-invariant inner product $\langle -,- \rangle_G$ and find a $T \in GL(n,\mathbb{C})$, such that $\langle x,y \rangle_G = \langle Tx,Ty \rangle$ with $\langle -,- \rangle$ being the canonical inner product. Then TGT^{-1} is contained in $U(n)$. We may of course assume that $\det T = 1$. Therefore

Any finite subgroup of $SL(n,\mathbb{C})$ *is conjugate in* $SL(n,\mathbb{C})$ *to a subgroup of* $SU(n)$.

Next using quaternions a 2:1 epimorphism $\rho: SU(2) \to SO(3)$ and an isomorphism $\tau: \mathbb{P} \to S^2$ will be described such that

(2) $\qquad \rho(g)\cdot\tau(z)=\tau(g\cdot z)$ for $z \in P$ and $g \in SU(2)$.

The complex 2×2 matrices of the form

$$q = \begin{pmatrix} a & -\bar{b} \\ b & \bar{a} \end{pmatrix}$$

are called *quaternions*. They form a real subalgebra $\mathbb{H}$ of the algebra of all 2×2 matrices of real dimension 4 with unit

$$e = \begin{pmatrix} 1 & 0 \\ 0 & 1 \end{pmatrix} .$$

The trace $\operatorname{tr} q = 2\cdot \operatorname{Re} a$ of a quaternion is real, the determinant $\det q = |a|^2+|b|^2$ is real and ≥ 0; $\det q = 0$ if and only if $q=0$. The transposed conjugate

$$q^* = \begin{pmatrix} \bar{a} & \bar{b} \\ -b & a \end{pmatrix}$$

of a quaternion q is also a quaternion. We have $q\,q^* = \det q\cdot e$. The algebra $\mathbb{H}$ is a skew field.

The algebra $\mathbb{H}$ is a Euclidean vector space with inner product

$$\langle p,q\rangle = \frac{1}{2}\cdot \operatorname{tr} pq \quad .$$

We have $|q|^2 = \langle q,q\rangle = \det q$.

The quaternions g with $|g|=1$ form the multiplicative group $SU(2)$. Obviously $SU(2)$ is homeomorphic to the 3-sphere $S^3 = \{(a,b)\in \mathbb{C}^2 : |a|^2+|b|^2 = 1\}$. Every $g\in SU(2)$ has two eigenvalues $e^{\pm i\varphi}$ with $0 \leq \varphi \leq \pi$. We have $\operatorname{tr} g = 2\cdot \cos\varphi$. The angle φ is the spherical distance between g and the unit e. Since $q \to g\cdot q$ is an orthogonal transformation the spherical distance between g and h in $SU(2)$ is the spherical distance between gh^{-1} and e. The orthogonal complement of $\mathbb{R}\cdot e$ in $\mathbb{H}$ is the three-dimensional vector space

$$V = \{v\in \mathbb{H}: \langle e,v\rangle = 0\} = \{v\in \mathbb{H}: \operatorname{tr} v = 0\} .$$

Every $g\in SU(2)$ determines the orthogonal transformation

(3) $$\rho(g): V \to V,\ v \to gvg^* .$$

Every g $SU(2)$ can be written as

(4) $$g = \cos\varphi\cdot e + \sin\varphi\cdot w \quad \text{with} \quad w\in V,\ |w| = 1 .$$

Vice versa, every element of the form (4) belongs to $SU(2)$

and $\operatorname{tr} g = 2\cos\varphi$. If g is written as in (4) then $\rho(g)$ is the rotation in V about the oriented axis through w with angle 2φ . Thus

$$\rho: SU(2) \to SO(V) \cong SO(3)$$

is a 2:1 epimorphism. The isomorphism $SO(V) \cong SO(3)$ depends on the choice of an orthonormal base of V , e.g.

$$e_1 = \begin{pmatrix} 1 & 0 \\ 0 & -1 \end{pmatrix}, \quad e_2 = \begin{pmatrix} 0 & 1 \\ 1 & 0 \end{pmatrix}, \quad e_3 = \begin{pmatrix} 0 & -i \\ i & 0 \end{pmatrix}.$$

In order to define $\tau: \mathbb{P} \to S^2$ we choose the base points $p = [1:0] \in \mathbb{P}$ and $q = e_1 \in S^2 = \{v \in V: \langle v,v\rangle = 1\}$. For every $z \in \mathbb{P}$ there is an $h \in SU(2)$ with $z = h \cdot p$. (Since the homogeneous coordinates (z_0, z_1) are determined only up to some factor $\lambda \in \mathbb{C}^\times$ we may assume $|z_0|^2 + |z_1|^2 = 1$. Then

$$h = \begin{pmatrix} z_0 & -\bar{z}_1 \\ z_1 & \bar{z}_0 \end{pmatrix}$$

transforms p into z) . The vector $\rho(h) \cdot q$ depends only on z and not on the choice of h : If also $z = g \cdot p$ then $h^{-1}g$ fixes p , hence $h^{-1}g = \begin{pmatrix} u & 0 \\ 0 & \bar{u} \end{pmatrix}$ for some u . Therefore $\rho(h^{-1}g)$ fixes q . Thus

$$\tau: \mathbb{P} \to S^2, \ \tau(z) = \rho(h) \cdot q \quad \text{for} \quad z = h \cdot p$$

is well defined. In order to prove injectivity assume that $z = h \cdot p$ and $z' = g \cdot p$ have the same image $\rho(h) \cdot q = \rho(g) \cdot q$. Then $\rho(h^{-1}g)$ fixes q and hence $h^{-1}g$ fixes p . In order to prove surjectivity we use that $SO(V)$ acts transitively on S^2 and that ρ is epimorphic. Hence for any $v \in S^2$ there is a $g \in SU(2)$ with $v = \rho(g) \cdot q = \tau(g \cdot p)$. The relation (2) follows directly from the definition of τ .

§3 Four-Dimensional Regular Solids

Since SU(2) is the unit sphere in the four-dimensional Euclidean space $\mathbb{H}$ of quaternions the convex hull ch G of any finite subgroup G < SU(2) is a convex polytope. If G is cyclic of order n then ch G is a regular n-gon. In all other cases ch G is four-dimensional as we shall see below.

The action of G on $\mathbb{H}$ by multiplication from the left embeds G as a subgroup into SO($\mathbb{H}$) . The elements of G are automorphisms of ch G . Since G acts transitively on itself, all elements of G are vertices of ch G .

We study the distribution of the elements of G among the primes of $\mathbb{H}$ which are perpendicular to the axis through Oe . and which meet SU(2). If L is such a prime, all elements of $L \cap SU(2)$ have the same spherical distance φ from e where $0 \le \varphi \le \pi$. Then L is called the *φ-prime* .

. The 90°-prime is the three-space V and ρ: SU(2) → SO(V) is the 2:1 epimorphism introduced in §2. Let $g \in SU(2)$. If $\rho(g)$ is an α-rotation for some $0 \le \alpha \le \pi$ then g lies in the $\alpha/2$- or in the $(\pi-\alpha/2)$-prime.

For a fixed $0 < \varphi < \pi$ the map $s = s_\varphi$: V → φ-prime, $s(v) = \cos\varphi \cdot e + \sin\varphi \cdot v$ is a $1:\sin\varphi$ similarity. If $|v| = 1$ then $s(v) \in SU(2)$ and $\rho(s(v)) \in SO(V)$ is the 2φ-rotation about the oriented axis through v . The restriction of ρ to $SU(2) \cap \varphi$-prime is injective if $\varphi \neq 90^o$.

Using these general observations the following distribution of the elements of the finite subgroup G < SU(2) among the φ-primes is obtained. Since $G \cap (\pi-\varphi)$-prime = $-(G \cap \varphi$-prime) it suffices to consider $0 \le \varphi \le \pi/2$. The figures 1-6 show two-dimensional sections of $\mathbb{H}$. The circle line represents SU(2) , the vertical lines represent those φ-primes which contain elements of G . The number of these elements is given at the bottom of the lines. Later we shall describe the configuration '$G \cap \varphi$-prime' in some interesting cases.

Figure 1-6:
The distribution of the elements of G among the φ-primes:

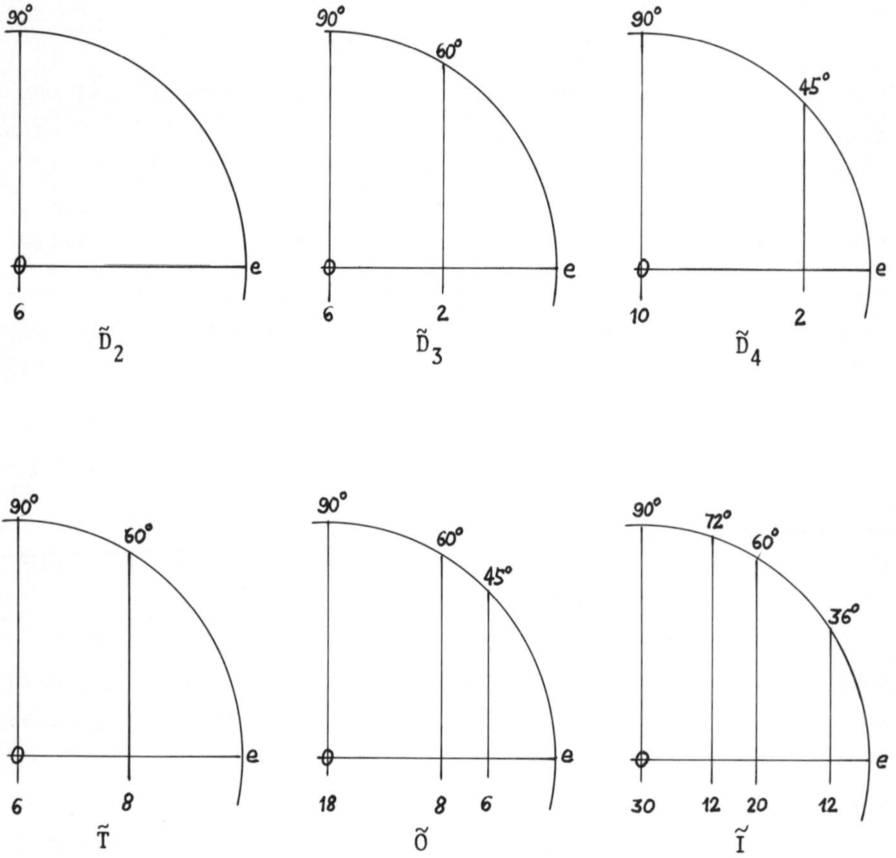

There are three regular solids among the ch G. The precise result is the following

<u>Theorem</u>: (a) *The convex hull of the binary dihedral group* $\tilde{D}_2$ *is the 4-octahedron* $\{3,3,4\}$ *. It has* 16 *tetrahedra as facets,* 32 *triangles as 2-faces,* 24 *edges, and* 8 *vertices.*

(b) *The convex hull of the binary tetrahedral group* $\tilde{T}$ *is the regular solid* $\{3,4,3\}$ *. It has* 24 *octahedra as facets,* 96 *triangles as 2-faces,* 96 *edges and* 24 *vertices.*

(c) *The convex hull of the binary icosahedral group* $\tilde{I}$ *is the regular solid* $\{3,3,5\}$ *. It has* 600 *tetrahedra as facets,* 1200 *triangles as 2-faces,* 720 *edges, and* 120 *vertices.*

(d) *For any other non-cyclic finite subgroup* $G < SU(2)$ *the convex hull* ch G *is not regular.*

Proof: Let $\alpha > 0$ be the smallest angle such that $G \cap \alpha$-prime $\neq \emptyset$. If the convex hull ch G is regular, $\sin \alpha/2$ is the characteristic ratio of ch G, and $Q = \text{ch}(G \cap \alpha\text{-prime})$ is the excellent vertex figure at e , see I §3. The comparism of $\sin \alpha/2$ and Q as given by the figures 1-6 above with the restricted possibilities of I §4 shows: At most ch $\tilde{D}_2$, ch T, and ch $\tilde{I}$ have a chance to be regular. For $\tilde{D}_n, n>3$, Q cannot be a vertex figure, and for $\tilde{O}$ $\sin 22{,}5^{\circ}$ does not occur as characteristic ratio of a regular solid. This proves (d). We claim for $G \in \{\tilde{D}_2, \tilde{T}, \tilde{I}\}$:

(I) $Q=\{3,4\}$ for $\tilde{D}_2$, $Q=\{4,3\}$ for $\tilde{T}$, and $Q=\{3,5\}$ for $\tilde{I}$.

(II) Every line joining e to any other $g \in G$ intersects the α-prime in Q .

(III) If for some $f \in O(\mathbb{H})$ the vertex figure Q is f-invariant, then the finite (eventually empty) set $G \cap \varphi$-prime is f-invariant for every angle $0 \leq \varphi \leq \pi$.

Then Q is a regular vertex figure at e and ch G is regular, because the criteria (i)-(iii) of I §3 are fullfilled. The number $n_o = \# G$ of vertices of ch G yields the numbers n_k of k-faces as stated in part (a)-(c) of the theorem: For each k-face we know the number q_k of its vertices, we know the number r_k of k-faces containing e , namely r_k=number of (k-1)-faces of Q . Then $n_k = n_o q_k / r_k$.

Proof of (I): The six points of $\tilde{D}_2 \cap V$ are easily seen to be the vertices of a regular octahedron. For $G=\tilde{T}$ and $=\tilde{I}$ let $\Delta \subset V$ be the regular solid which belongs to $\rho(G) \subset SO(V)$ and has $V \cap SU(2)$ as circum-sphere. For $G=\tilde{T}$ this Δ is a tetrahedron. We form the stella octangula $\Delta \cup (-\Delta)$. Its convex hull is a cube and $Q=s_\alpha(C)$ is also a cube. For $G=\tilde{I}$ our Δ is an icosahedron and so is $Q=s_\alpha(\Delta)$.

Proof of (II): This is obvious for $G=\tilde{D}_2$. For $G=\tilde{T}$ and $=\tilde{I}$ we caculate the in-radius $r=r_2$ of Q , i.e. the distance between the centre O of Q and the centre C of a face of

Q . Let R denote the circum-radius of Q . If Q={p,q} then

$$(1) \qquad (r/R)^2 = 1 - \frac{s(Q)^2}{\sin^2 \pi/p} \quad ,$$

where s(Q)=1/2R denotes the characteristic ratio, see I§3 . The formula (1) follows from two rectangular triangles, see figure 7. Here V denotes a vertex and M a mid-edge point.

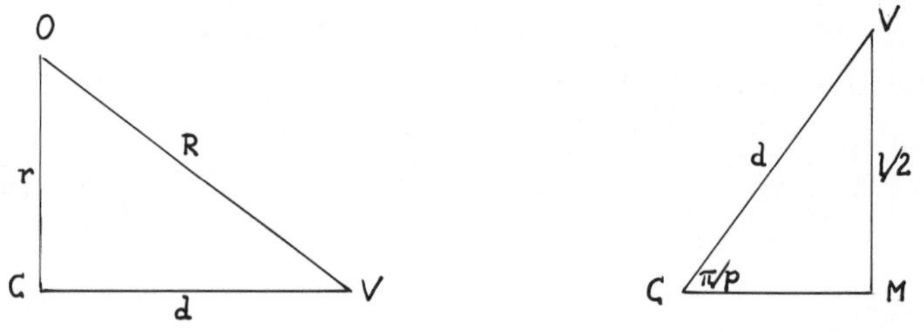

Figure 7

This implies $(r/R)^2=1/3$ for the cube and $(r/R)^2=(5+2\sqrt{5})/15$ for the icosahedron. The circumradius of Q is $R=\sin\alpha$, hence $R=\sin 60^o=\sqrt{3}/2$ for T where Q is a cube and $R=\sin 36^o=\sqrt{(5-\sqrt{5})/8}$ for I where Q is an icosahedron. Figures 8 and 9 show these radii r and complete the proof of (II) .

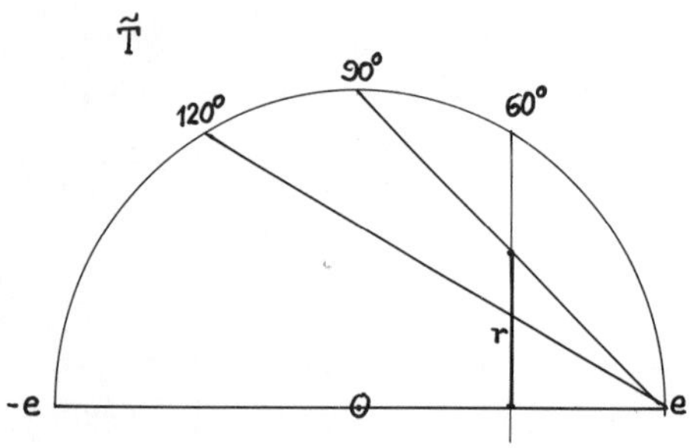

r=0.5

Figure 8

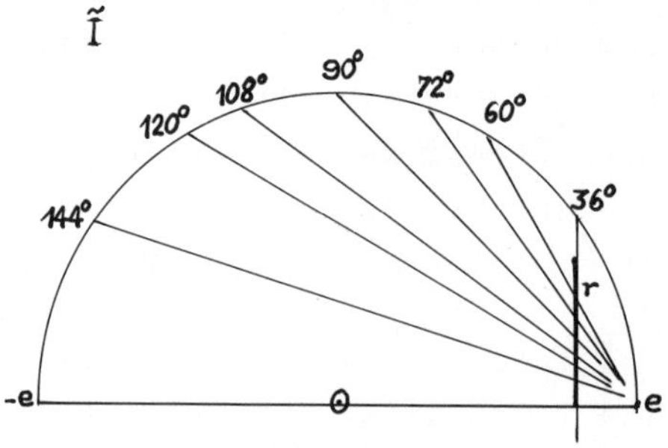

r=0.4671

Figure 9

Proof of (III): It suffices to consider the angles $0 \le \varphi \le \pi/2$. We have f(e)=e . This proves the case $\varphi=0$. It suffices for $\tilde{D}_2$. For $\tilde{T}$ the φ-prime for one more angle $\varphi=\pi/2$, i.e.

$\tilde{T} \cap V$ must be considered besides Q. Since the cube $Q=s_\alpha(C)$ is f-invariant, C itself is f-invariant, hence also the set of the six face centres of C, radially projected onto the unit sphere $SU(2) \cap V$. This set (the vertices of an octahedron) is $\tilde{T} \cap V$. For $\tilde{I}$ the angles $\varphi \in \{60^o, 72^o, 90^o\}$ must be considered besides $\alpha=36^o$. Since the icosahedron $s_\alpha(\Delta)$ is f-invariant, Δ itself is f-invariant. Now each $\tilde{I} \cap \varphi$-prime is obtained from Δ by an f-invariant construction: Let Θ be the dual dodecahedron of Δ, whose vertices are the face centres of Δ radially projected onto the unit sphere $SU(2) \cap V$. Then $\tilde{I} \cap 60^o$-prime consists of the vertices of $s_\varphi(\Theta)$ for $\varphi=60^o$. Further, $\tilde{I} \cap 72^o$-prime consists of the vertices of $s_\varphi(\Delta)$ for $\varphi=72^o$. Finally, the set of mid-edge points of Δ, radially projected onto the unit sphere $SU(2) \cap V$ is the set $\tilde{I} \cap V$. (Its convex hull is the semi-regular icosidodecahedron, see figure 10.)

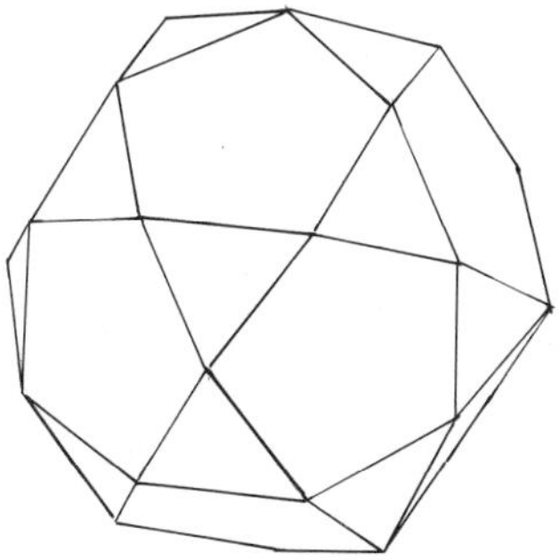

Figure 10

§4 The Orbit Spaces S^3/G of the Finite Subgroups G of $SU(2)$

Every finite subgroup $G < SU(2)$ acts on the 3-sphere $SU(2)$ by multiplication from left. We introduce some general notions in order to study such actions.

Let X be a topological space and let G be a group. An *action* of G on X from the left assigns to every $g \in G$ a continuous map $X \to X$, $x \mapsto gx$, with the properties $1 \cdot x = x$ and $(hg)x = h(gx)$ for all $x \in X$ and $g, h \in G$. The subset $Gx = \{gx : g \in G\} \subset X$ is called the *orbit* through x. The total space X is the disjoint union of all orbits. The set of all orbits is denoted by X/G ; some authors prefer $G \backslash X$ in order to emphazise that G acts from the left. There is the canonical projection

$$p: X \to X/G \ , \ p(x) = Gx \ .$$

The orbit set X/G is made into a topological space: A subset $U \subset X/G$ is open if and only if $p^{-1}(U) \subset X$ is open. The projection p is a continuous and open map.

In the following, the group G is assumed to be *finite*. Then p is also closed. The orbit space X/G is a [compact] Hausdorff space if X is a [compact] Hausdorff space, see also IV§2. If G acts ***freely***, i.e.if $gx = x$ for some $g \in G$ and $x \in X$ implies $g = 1$, then p is a covering projection: Every $x \in X$ has a neighbourhood U such that $p^{-1}(p(U)) = \dot{\bigcup}_{g \in G} gU$ (disjoint union), and p maps each gU homeomorphically onto $p(U)$. The group G is the group of *covering* (or *deck*) transformations of X over X/G , i.e. every homeomorphism f of X with $p \circ f = p$ is the action of some $g \in G$ on X .

If X is a differentiable manifold and if G acts freely via differentiable maps, the orbit space X/G inherits a unique differentiable structure so that p becomes locally diffeomorphic. If in addition X is oriented and every map $X \to X$

$x \mapsto gx$, is orientation preserving then X/G inherits a unique orientation so that p is orientation preserving.

Theorem: *Let* $G < SU(2)$ *be a finite subgroup which acts on* $S^3 = SU(2)$ *by multiplication (from left). The orbit space* S^3/G *is a compact oriented 3-manifold. The canonical projection* $p: S^3 \to S^3/G$ *is the universal covering. The fundamental group* π_1 *and the integral homology groups* H_i *of* S^3/G *are the following ones:*

π_1	H_0	H_1	H_2	H_3
G	$\mathbb{Z}$	$\mathcal{A}G$	0	$\mathbb{Z}$

$\mathcal{A}G$=abelianized group G, see table 2 at the end of the following §5.

Proof: The 3-sphere S^3 is a differentiable compact oriented 3-manifold, the group G acts differentiably, orientation preserving, and freely. Therefore S^3/G is also a differentiable, compact oriented 3-manifold. The projection p is the universal covering because S^3 is simply connected. Therefore the group G of covering transformations is canonically isomorphic to the fundamental group $\pi_1(S^3/G)$. Since S^3/G is path connected, $H_0=\mathbb{Z}$. A general result for arbitrary topological spaces says $H_1(X) = \mathcal{A}\pi_1(X)$, thus $H_1=\mathcal{A}G$. By Poincaré duality $H^1=H_2$ for every compact oriented 3-manifold. The Universal Coefficient Theorem, i.e. the exact sequence for arbitrary spaces $0 \to \mathrm{Ext}(H_{q-1},\mathbb{Z}) \to H^q \to \mathrm{Hom}\,(H_q,\mathbb{Z}) \to 0$ yields $H^1=0$ because H_1 is finite and $H_0=\mathbb{Z}$.

A suitable fundamental domain may help to imagine the orbit manifolds S^3/G . In general, a subset $D \subset X$ is called a *fundamental domain* for the action of the finite group G on X if it satisfies:

(1) For every $x \in X$ there is a $g \in G$ with $gx \in D$.

(2) If for some $g \in G$ both $x, gx \in \mathrm{int}\, D$ (interior), then $g=1$.

(3) $D =$ closure of $\mathrm{int}\, D$.

Then $X/G = p(D)$, the projection p maps $\mathrm{int}\, D$ homeomorphically onto $p(\mathrm{int}\, D)$ and $p(\mathrm{int}\, D)$ is dense in X/G . Thus

X/G is obtained from D by identifying points of the boundary $\partial D = D \setminus \text{int } D$ only.

Let $P = (\text{ch } G)^o$ be the polar polytope of the convex hull of G with respect to the 3-sphere $SU(2)$. The multiplication of G from left leaves ch G and hence P invariant. The radial projection maps the boundary ∂P = union of all facets of P homeomorphically onto $S^3 = SU(2)$. This homeomorphism is G-invariant therefore $\partial P/G$ is homeomorphic to S^3/G.

The centres of the facets of P are the elements $g \in G$. The facet with centre g is denoted by D_g. The multiplication from left by the element $h \in G$ maps D_g isomorphically onto D_{hg}. Therefore each D_g is a fundamental domain for the action of G on ∂P. For simplicity we choose $D = D_e$, where e is the unit element. This facet D is surrounded by the facets D_g for all $g \in G$ which are connected to e by an edge of ch G. The set of these g will be denoted by st(e), the star of e. If ch G is regular, st(e) consists of the vertices of the excellent vertex figure of ch G at e, but in the other cases st(e) does not lie in *one* prime.

For every $g \in \text{st}(e)$ there is the common two-face $F_g = D \cap D_g$, and $\partial D = \bigcup_{g \in \text{st}(e)} F_g$. The orbit space $\partial P/G$ is obtained from D by identifying every $F_{g^{-1}}$ with F_g via the multiplication with g from left. Geometrically this isomorphism $g: F_{g^{-1}} \to F_g$ is a φ-screw, i.e. a translation along the vector from the centre of $F_{g^{-1}}$ to the centre of F_g (this vector is perpendicular to both F_g and $F_{g^{-1}}$, which lie in parallel two-planes) combined with a φ-rotation about the axis through this vector, where φ is the spherical distance from g to e.

Let $\alpha > 0$ be the smallest angle such that $G \cap \alpha\text{-prime} \neq 0$. The intersection $Q = (\text{ch } G) \cap \alpha$-prime is a vertex figure of ch G at e with centre $c = \cos\alpha \cdot e$. There is a one-to-one correspondence between the star st(e) and the set $V(Q)$ of vertices of Q: The edge from e to $g \in \text{st}(e)$ intersects the α-prime at $v_g \in V(Q)$. Let Q^c be a polar of Q with respect to some 2-sphere about c in the α-prime. Every vertex v_g of Q determines the 2-face v_g^Δ of Q^c , see I§2. According to the proposition at the end of I§2 the facet D of $P=(\operatorname{ch} G)^o$ is similar to Q^c . The similarity maps the 2-face v_g^Δ of Q^c onto the 2-face F_g of D .

In the three cases $G=\tilde{D}_2$, $=\tilde{T}$, and $=\tilde{I}$, where ch G is regular, the star st(e)=V(Q) . Both Q and D are mutually dual regular solids. The orbit space S^3/G is obtained from D by the identification of opposite 2-faces of D by α-screws.

G	D_2	T	I
Q	{3,4}	{4,3}	{3,5}
D	{4,3}	{3,4}	{5,3}
α	90°	60°	36°

The figures 1-3 show the fundamental domains D of S^3/G for $G=\tilde{D}_2$, $=\tilde{T}$, and $=\tilde{I}$. Edges with the same letter are identified by the screws.

If $G=\tilde{D}_n$ with $n \geq 3$ or $G=\tilde{O}$ the elements g of st(e) partly belong to the α-prime (then $g=v_g$) and partly to one other β-prime. Therefore Q has two types of vertices, called α- and β-vertices, and D has two types of faces, called α- and β-faces.

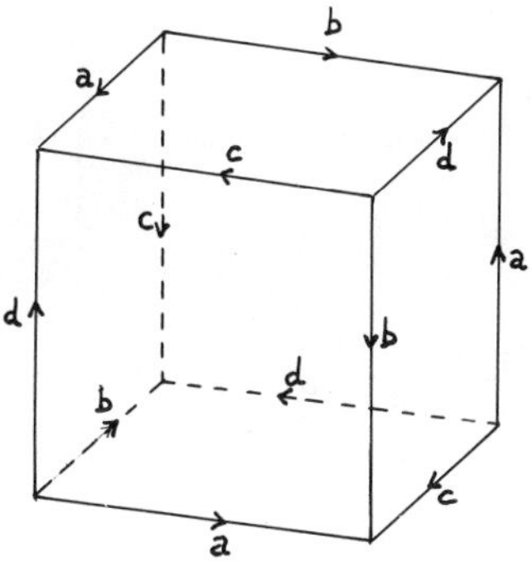

Figure 1

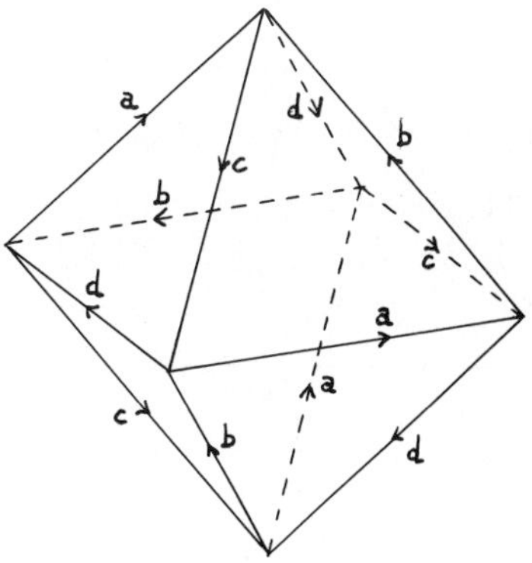

Figure 2

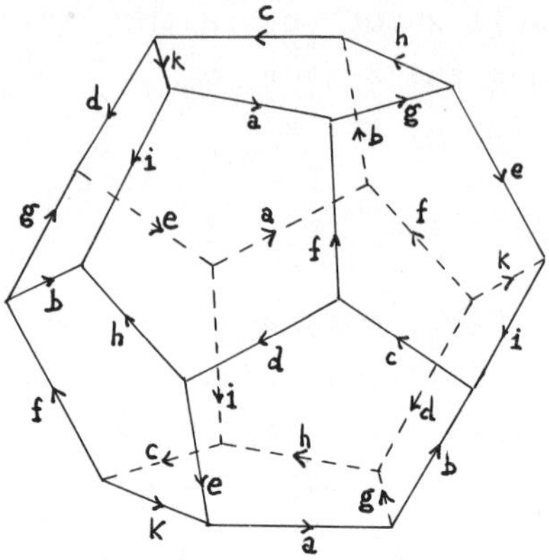

Figure 3

For $G=\tilde{D}_n$ $(n\geq 3)$ we have $\alpha = 180^o/n$ and $\beta = 90^o$. The vertex figure Q is a double pyramid over a 2n-gon, see figure 4 for n=3. The two apices are α-vertices, the vertices of the equatorial 2n-gon are β-vertices. The polar D of Q is a prism over a 2n-gon. Its top and bottom 2n-gons are α-faces, the lateral squares are β-faces. The figure 5 shows D for n=3. Edges which are identified by the α- and β-screws are denoted by the same letter.

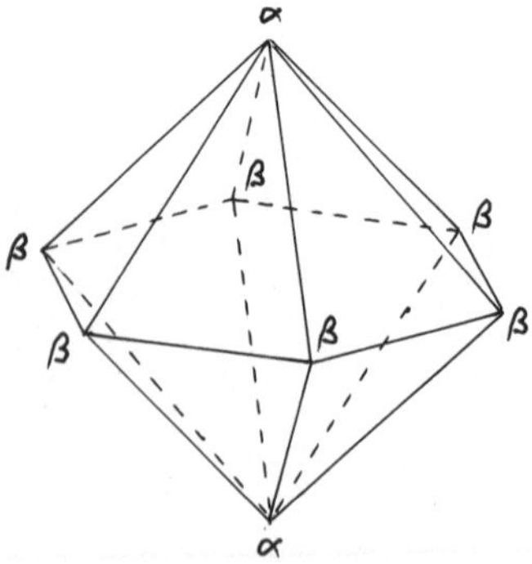

Figure 4

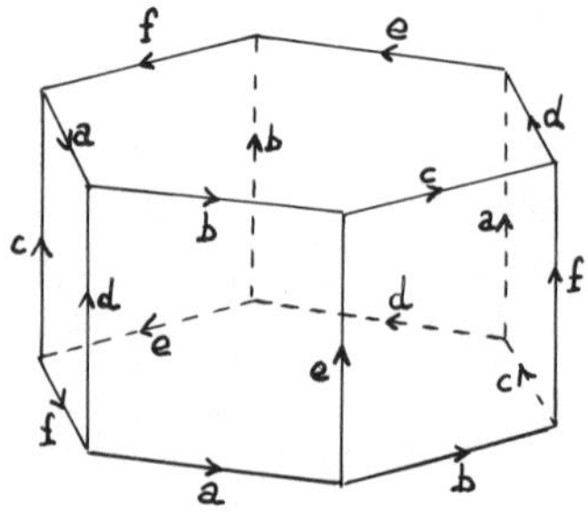

Figure 5

For $G=\tilde{O}$ we have $\alpha=45^{o}$ and $\beta=60^{o}$. The six α-vertices are the vertices of an octahedron Ω, the eight β-vertices are the vertices of a cube W, which is dual to Ω. Figure 6 shows the circumradii $R_{\Omega} = \sqrt{2}/2 = 0.7071$ and $R_W = \sqrt{3}(2-\sqrt{2})/2 = 0.5073$. Observe that R_W is bigger than the in-radius $r_{\Omega} = 1/\sqrt{6} = 0.4082$, so that W is not contained in Ω.

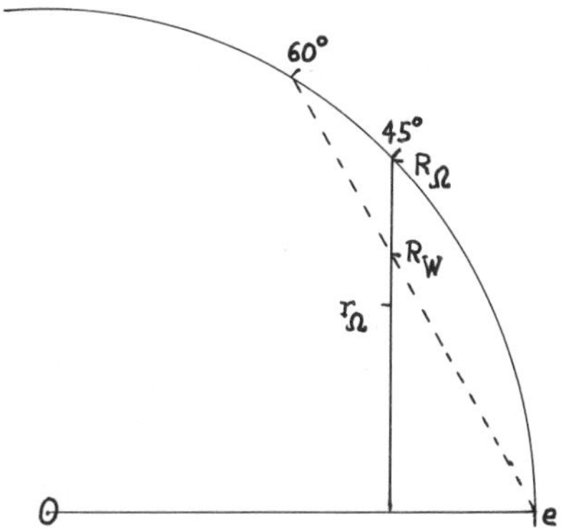

Figure 6

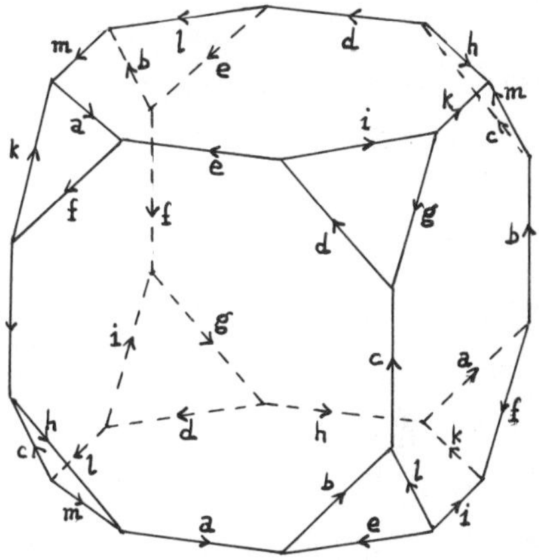

Figure 7

The vertex figure is $Q = ch(\Omega \cup W)$. The facet D is similar to the polar $Q^c = (ch\ \Omega)^c \cap (ch\ W)^c$ which is the intersection of the cube $(ch\ \Omega)^c$ and its dual octahedron $(ch\ W)^c$ (a truncated cube), see figure 7. The α-faces of D are regular octagons, the β-faces are equilateral triangles. Again opposite α-faces are identified by α-screws and opposite β-faces by β-screws in order to obtain the orbit space $S^3/\tilde{O}$.

Remark:
The fundamental domain D together with the identification of its opposite faces yields a cell decomposition of the orbit space S^3/G with the following numbers n_i of i-dimensional cells:

G	n_0	n_1	n_2	n_3
D_n	n	2n	n+1	1
T	1	4	4	1
O	6	12	7	1
I	5	10	6	1

Observe that the Euler-Poincaré characteristic $n_o - n_1 + n_2 - n_3 = 0$ in all cases as it must be the case with an compact oriented 3-manifold. These cell decompositions can (and have) also been used in order to determine the fundamental group and the homology of S^3/G. But in contrast to the methods of algebraic topology which we employed above this requires some calculations, see Seifert-Threlfall (2) §61, §62 and DuVal 4.32.

§5 Generators and Relations for the Finite Subgroups of SL(2,ℂ)

Let $G < SL(2,\mathbb{C})$ be a binary finite subgroup and let $\rho: G \to \Gamma$ be the 2:1 epimorphism onto the corresponding finite subgroup $\Gamma < SO(3)$.

<u>Lemma 1</u>: *If* $\gamma \in \Gamma$ *has order* r, *there is a* $g \in G$ *with* $\rho(g)=\gamma$ *and* $g^r=-e$. *Here* e *denotes the unit.*

Proof: Remember that $-e$ is the only element of order 2 in $SL(2,\mathbb{C})$, see §1 Lemma 2. There is a g with $\rho(g)=\gamma$ and $g^r=\pm e$. Assume $g^r=e$. If r is odd, g is replaced by $-g$. If $r=2s$ is even, we obtain a contradiction: $g^{2s}=e$ implies $g^s=\pm e$. Therefore $\gamma^s=1$, but γ has order $2s$.

The presentation of the finite subgroups $G < SL(2,\mathbb{C})$ by generators and relations is obtained from the presentation of the corresponding finite subgroups $\Gamma < SO(3)$, see I§8. If G is not cyclic, Γ has two generators β and γ. Using Lemma 1 we choose two elements $b,c \in G$ such that $\beta=\rho(b)$, $\gamma=\rho(c)$, and such that the orders of b and c are twice the order of β and γ respectively. These generators yield the following presentation

type of $G<SL(2,\mathbb{C})$	generators	relations
cyclic of order n	b	$b^n=1$
binary q-dihedral	b,c	$b^2=c^q=(bc)^2$
binary tetrahedral	b,c	$b^3=c^3=(bc)^2$
binary octahedral	b,c	$b^3=c^4=(bc)^2$
binary icosahedral	b,c	$b^3=c^5=(bc)^2$

Table 1

The proof consists of three parts: (a) The relations hold true. (b) The generators b and c suffice. (c) The relations suffice.

Part (a): Let q and r be the orders of β and γ respectively. Then b and c satisfy $b^q=c^r=-e$. Thus it suffices to show that $(bc)^2=-e$. Since $(\beta\gamma)^2=1$ we have $(bc)^2=\pm e$. The assumption $(bc)^2=e$ would yield $bc=\pm e$, hence $\beta\gamma=1$.

Part (b): Let G' be the subgroup of G which is generated by b and c . Then $-e=(bc)^2\in G'$, further ρ maps G' epimorphically onto Γ . Hence $G'=G$.

Part (c): Let G^* be the group which is presented according to the table. The element $v=(bc)^2$ belongs to the centre of G^* . We show that $v^2=1$: If $b^2=v$ we have $c=b^{-1}c^{-1}b$ whence $v=c^q=bc^{-q}b^{-1}=bv^{-1}b^{-1}=v^{-1}$. If $b^3=v$ we have $c=b\cdot bc^{-1}\cdot b^{-1}$ whence $v=c^r=b(bc^{-1})^r b^{-1}$ thus

(1) $$v=b^{-1}vb=(bc^{-1})^r \quad .$$

By substitution of $b=c^{-1}b^{-1}c^{r-1}$ we obtain $v=c^{-1}(b^{-1}c^{r-3})^r c$ thus

(2) $$v=(b^{-1}c^{r-3})^r \quad .$$

The conjugation $b..b^{-1}$ transforms this into

(3) $$v=(c^{r-3}b^{-1})^r \quad .$$

r=3: $v \overset{(2)}{=} b^{-3}=v^{-1}$.

r=4: $v \overset{(3)}{=} (cb^{-1})^4=(bc^{-1})^{-4} \overset{(1)}{=} v^{-1}$.

r=5: $v=(b^{-1}c^2)^5$ by (3). We substitute $c=b^2c^{-1}b^{-1}$ and obtain $v=(bc^{-1}bc^{-1}b^{-1})^5=bc(c^{-2}b)^5c^{-1}b^{-1} \overset{(2)}{=} bc\cdot v^{-1}c^{-1}b^{-1}=v^{-1}$.

There is an epimorphism $h: G^* \to G$ because of (a) and (b). Comparing the presentation of G^* with the presentation of $\Gamma=\rho(G)$ as in I§8 we see that the kernel of $\rho\circ h: G^* \to G \to \Gamma$ is generated by v. Thus the kernel of h is generated by some power v^s. Since $h(v)=-e$ and $v^2=1$ the exponent is $s=0$, i.e. h is an isomorphism.

<u>Remark</u>: The reader may perhaps wish to deduce $v^2=1$ from the given relations $b^q=c^r=(bc)^2$ by a general argument which works for all q and r. This is impossible because for all other q,r the element v has infinite order. Here is the sketch of a geometric proof. For more details see the proof of Lemma 3.1 in Milnor (2). The numbers q and r which occur with the finite subgroups of $SL(2,\mathbb{C})$ are exactly those which satisfy $1/2+1/q+1/r > 1$. In all other cases $1/2+1/q+1/r \leq 1$. Hence in the other cases there is a Euclidean or hyperbolic triangle with angles $\pi/2, \pi/q$, and π/r, see the figure.

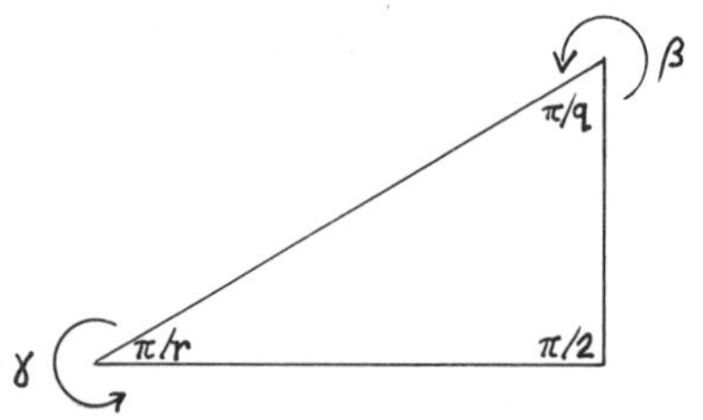

Let β and γ denote the rotations as in the figure through angles $2\pi/q$ and $2\pi/r$ respectively. Then $\beta^q=\gamma^r=(\beta\gamma)^2=\mathrm{id}$. Let B denote the group of all orientation preserving isometries of the plane, let $\tilde{B}$ denote the universal covering group. The canonical projection $\rho: \tilde{B} \to B$ has an infinite cyclic kernel C, in contrast to the 'spherical' case $\tilde{B}=SU(2)$ and $B=SO(3)$ where C has order 2. If the elements $b,c \in \tilde{B}$ with $\rho(b)=\beta$ and $\rho(c)=\gamma$ are suitably chosen they satisfy $b^q=c^r=(bc)^2=z$, and z generates C. Thus there is a homomorphism from the abstract group generated by b,c, with relations $b^q=c^r=(bc)^2$ into $\tilde{B}$ such that the image z of $v=b^q$ has infinite order.

The presentations of G in table 1 easily yield the abelianized groups $\mathcal{A}G$, see table 2. The epimorphism $\rho: G \to \Gamma < SO(3)$

induces an epimorphism $\bar{\rho}: \mathcal{A}G \to \mathcal{A}\Gamma$ It is 2:1 if G is cyclic of even order or binary q-dihedral with q odd. In all other cases $\bar{\rho}$ is an isomorphism. (The $\mathcal{A}\Gamma$'s are easily obtained from I§7 or §8).

type of $G < SL(2,\mathbb{C})$	the abelianized group G
cyclic of order n	$\mathcal{A}G = G$
binary q-dihedral	q even: Four group generated by b and c
	q odd: cyclic of order 4, generated by b with $c=b^2$
binary tetrahedral	cyclic of order 3, generated by c with $b=c^{-1}$
binary octahedral	cyclic of order 2, generated by c with b=1
binary icosahedral	{1}

Table 2

§6 Invariant Divisors and Semi-Invariant Forms

From this section on we shall study the rings of G-invariant polynomials in two variables for the finite subgroups $G < SL(2,\mathbb{C})$. We begin with a few general definitions and observations.

A subgroup $G < GL(n,\mathbb{C})$ acts from the right on the $\mathbb{C}$-algebra $S = \mathbb{C}[z_1,\dots,z_n]$ of all polynomials f in n variables by means of

$$(f\gamma)(z)=f(\gamma z) \quad \text{for } \gamma \in G \text{ and } z \in \mathbb{C}^n .$$

The polynomial f is called G-*invariant* if $f\gamma=f$ for every

$\gamma \in G$. The G-invariant polynomials form a subalgebra S^G. In order to study S^G we may replace G by a conjugate subgroup $H=\alpha G\alpha^{-1}$ because $S^G \to S^H$, $f \to f\alpha$, is an isomorphism. The action of G respects the additive decomposition of a polynomial f into its homogeneous terms. Particularly f is G-invariant if and only if its homogeneous terms are G-invariant. Therefore we first restrict attention to the action of G on *forms* (= homogeneous polynomials).

The *Fundamental Theorem of Algebra* says: Every complex polynomial in one variable can uniquely be written as

$$\varphi(z)=c(z-z_1)^{r_1}\dots(z-z_q)^{r_q}$$

where $z_1,\dots,z_q$ are the different zeros of φ which have orders (multiplicities) $r_1,\dots,r_q$ respectively. There is a close relationship between forms in two variables and polynomials in one variable: The form $f(z,w)$ of degree n determines the polynomial $\varphi(z)=f(z,1)$ of degree $\leq n$ and $f(z,w)=w^n\varphi(z/w)$. Thus the Fundamental Theorem implies for forms:

Every form in two variables can be written as

$$f(z,w) = (b_1z-a_1w)^{r_1}\dots(b_qz-a_qw)^{r_q}$$

where $[a_1:b_1]$, ..., $[a_q:b_q] \in \mathbb{P}$ (the projective line, see §1) are the different zeros of f which have orders $r_1,\dots,r_q$ respectively $(r_\nu \in \mathbb{N},\ r_\nu \neq 0)$. The form f is uniquely determined up to a factor in $\mathbb{C}^\times$ by its zeros and their orders. The finite set of zeros together with the orders can be prescribed arbitrarily. For a better formulation the notion of "divisor" will be used:

A *divisor* on $\mathbb{P}$ is a function $D\colon \mathbb{P} \to \mathbb{Z}$ such that $D(x)=0$ for all but finitely many x, The divisor D is called *positive*, if $D(x) \geq 0$ for every x. The form f determines the positive divisor $D_f(x)$ = order of x as zero of f (in particular $D_f(x)=0$ if x is not a zero). For every positive divisor D there is a form f, uniquely determined up to a factor in $\mathbb{C}^\times$, with $D_f=D$. Divisors can be added. This corresponds to the multiplication of forms: $D_{fg}=D_f+D_g$. All divisors (not only the positive ones) form an abelian group $\mathcal{D}$. The *degree* $\deg D = \sum_{x\in\mathbb{P}} D(x)$ is a homomorphism $\mathcal{D} \to \mathbb{Z}$. Obviously $\deg D_f = \deg f$.

A finite subgroup $\Gamma < SO(3)$ acts linearly on $\mathbb{P}$ from the left, see §1. Hence it acts on the group of divisors from the right:

$$(D\gamma)(x)=D(\gamma x) \quad \text{for } \gamma\in\Gamma \text{ and } x\in\mathbb{P}.$$

The divisor D is called Γ-*invariant* if $D\gamma=D$ for every $\gamma\in\Gamma$. The invariant divisors form a subgroup. Let $\Sigma\subset\mathbb{P}$ be an orbit of the Γ-action. It determines the positive Γ-invariant divisor

$$D(x)=1 \text{ if } x\in\Sigma, \quad D(x)=0 \text{ else}.$$

This D is called the *simple* divisor of Σ. It will often be convenient to denote the simple divisor of Σ by the same letter Σ.

Proposition 1: *Every positive invariant divisor can uniquely be written as finite sum*

$$D=r_1\Sigma_1+\ldots+r_k\Sigma_k \tag{1}$$

of different simple divisors Σ_i *with coefficients* $r_i\in\mathbb{N}$, $r_i>0$.

Proof: Let Σ_i be an orbit. Then D has the same value r_i at all points of Σ_i. This implies (1).

Let $G < SL(2,\mathbb{C})$ be a finite subgroup. Up to conjugacy we may assume $G < SU(2)$. Let $\rho\colon G\to\Gamma < SO(3)$ be the epimorphism, see §1. For a form f with divisor D_f we have $D_{f\gamma}=(D_f)\rho(\gamma)$ for $\gamma\in G$, in particular: If f is G-invariant then D_f is Γ-invariant. But the converse is not quite true: If D_f is Γ-invariant then $f\gamma$ equals f up to a factor in $\mathbb{C}^\times$. This motivates the following definition.

A form f is called G-*semi-invariant*, if for each $\gamma\in G$ there is a $\chi(\gamma)\in\mathbb{C}^\times$ such that

$$f\gamma=\chi(\gamma)f \quad \text{i.e.} \quad f(\gamma z)=\chi(\gamma)f(z) \quad \text{for } z\in\mathbb{C}^2.$$

If $f\neq 0$ the factor $\chi(\gamma)$ is uniquely determined by f and

γ . The mapping $\chi: G \to \mathbb{C}^{\times}$ is a homomorphism. It is called the *character* of f . In order to emphazise f we write χ_f . If $N = \# G$ the values of χ are N-th roots of unity.

The form f is G-semi-invariant if and only if its divisor D_f is Γ-invariant. The character of f depends only on D_f . Hence for every positive Γ-invariant divisor there is a well defined character $\chi_D: G \to \mathbb{C}^{\times}$. Obviously $\chi_{C+D} = \chi_C \cdot \chi_D$ for two positive divisors. Therefore proposition 1 implies that the character of an arbitrary invariant divisor can be calculated as soon as the characters of the simple divisors are known. We shall see in a moment that it suffices to know the characters of the two or three *exceptional divisors*, i.e. the simple divisors of the two or three exceptional orbits.

For a point $x \in \mathbb{P}$ let x also denote the divisor which has value 1 at x and value 0 else. The positive invariant divisor $D = \sum_{\gamma \in \Gamma} \gamma x$ of degree $N = \# \Gamma$ is called the *principal divisor* through x . If Σ is the orbit through x then $D = n\Sigma$ where $n = \# \Gamma_x$ is the order of the isotropy subgroup of x . For every $\gamma \in G$ the character value $\chi_D(\gamma)$ depends continuously on x . Since $\mathbb{P}$ is connected and $\chi_D(\gamma)$ is a $(\# G)$-th root of unity, $\chi_D(\gamma)$ does not depend on D . We have proved,

Proposition 2: *All principal divisors have the same character. If* Σ *is an exceptional* Γ*-orbit then the common character of all principal divisors is* $\chi = \chi_{\Sigma}^{n}$ *with* $n = \# \Gamma / \# \Sigma$.

§7 The Characters of the Invariant Divisors

As in the preceding §6 let $G < SU(2)$ be finite and $\Gamma = \rho(G) < SO(3)$. In order to obtain all G-invariant polynomials we proceed as follows: First the characters of the exceptional divisors are determined. Then we know the characters of all simple divisors because of §6, proposition 2. We can calculate the character of every invariant divisor using §6, proposition 1. In particular, we obtain those invariant divisors which have trivial character. Their corresponding forms are exactly the G-invariant forms, i.e. the homogeneous components of the G-invariant polynomials.

Cyclic Groups: *Let the cyclic group* G *of order* n *be generated by the element* b *. There are two exceptional divisors* P *and* Q *of* Γ *. They belong to the two fixed points. The character values* $\chi_P(b)$ *and* $\chi_Q(b)$ *are primitive* n*-th roots of unity with* $\chi_P \cdot \chi_Q = 1$ *. The common character* χ *of the principal divisors is determined by the value* $\chi(b)=1$ *for odd* n *and* $\chi(b)=-1$ *for even* n *.*

Proof: We may assume that (after suitable conjugation)

$$b = \begin{pmatrix} \eta & 0 \\ 0 & \eta^{-1} \end{pmatrix} \quad \text{with} \quad \eta = e^{2\pi i/n} \quad .$$

Then $P=[0:1]$ and $Q=[1:0]$ are the fixed points and $f(z,w)=z$ and $g(z,w)=w$ are two forms with $D_f=P$ and $D_g=Q$. This yields $\chi_P(b)=\chi_f(b)=\eta$ and $\chi_Q(b)=\eta^{-1}$. If n is odd, $n=\#\Gamma$ and hence nP is principal so that $\chi(b)=\chi_P(b)^n=1$. If n is even, $n/2 = \#\Gamma$ so that $\chi(b)=\chi_P(b)^{n/2}=-1$. (The difference between even and odd n comes from §1, Lemma 3) .

Non-Cyclic Groups: We use the results which are summarized at the end of I§6. There are three exceptional divisors E, V, F which belong to the exceptional orbits of mid-edge points, vertices, and face centres, respectively. Since the values of the characters lie in an abelian group, every character $\chi: G \to \mathbb{C}^\times$ factorizes through the abelianized group $\mathcal{A}G$, see the table 2 at the end of §5. It suffices to determine the values $\chi(b)$ and $\chi(c)$ for the generators b, c of $\mathcal{A}G$. In the following table 1 χ denotes the common character of the principal divisors and θ is a primitive third root of unity.

type of $G < SL(2,\mathbb{C})$		χ_E	χ_V	χ_F	χ
binary q-dihedral, q odd	b	i	-i	-1	-1
binary q-dihedral, q even	b	1	-1	-1	1
	c	-1	-1	1	1
binary tetrahedral	c	1	θ	θ^2	1
binary octahedral	c	-1	-1	1	1
binary icosahedral	1	1	1	1	1

Table 1

Proof of table 1: The last line follows from the fact that the abelianized binary icosahedral group is trivial. Since all elements of $\mathcal{A}G$ have order ≤ 2 , if G is binary q-dihedral, with q even, or binary octahedral, the corresponding character values are ± 1 . The character values of the binary q-dihedral group for q odd are fourth roots of unity because $\mathcal{A}G$ is cyclic of order 4 . They are third roots of unity with the binary tetrahedral group because then $\mathcal{A}G$ is cyclic of order 3. In all cases $\chi = \chi_E^2$ because $2E$ is a principal orbit. After these general observations we show how the precise values of $\chi(b)$ are obtained for the binary q-dihedral group with q odd. The reader will then be able to calculate all other values of table 1 by the same method: The element $b \in G$ has order 4.

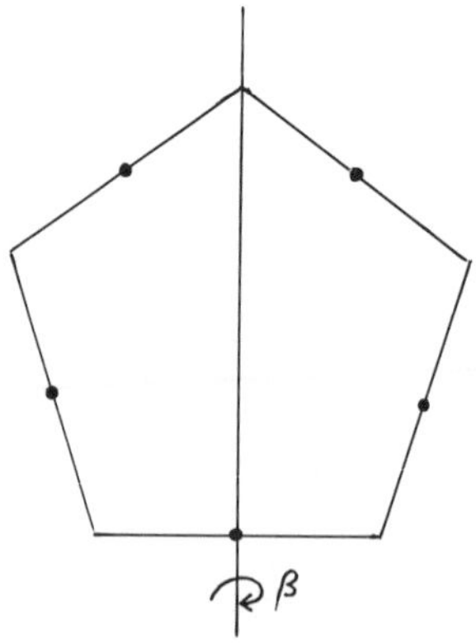

Its image $\beta = \rho(b) \in \Gamma$ is a π-rotation about a symmetry axis of the invariant q-gon, see the figure for $q=5$. The Γ-invariant orbit E of mid-edge points (the dots in the figure) consists of one fixed point of β and $(q-1)/2$ many principal β-orbits. We apply the result obtained above to the cyclic group of order 4 generated by b : The fixed point contributes the factor $\pm i$ to $\chi_E(b)$, each principal β-orbit contributes the factor -1 , hence $\chi_E(b) = \pm i$. If b is eventually replaced by b^{-1} we normalize to $\chi_E(b) = i$. The Γ-orbit of vertices consists of the other fixed point of β and $(q-1)/2$ many principal β-orbits, hence $\chi_V(b) = \chi_E(b)^{-1} = -i$. The Γ-orbit F is a principal β-orbit, thus $\chi_F(b) = -1$.

The orbit structure of the Γ-action on $\mathbb{P}$, see the summary at the end of I,§6, and the decomposition of invariant divisors into simple divisors according to proposition 1 in §6 imply that every positive invariant divisor can be written as

(1a) $D=rP+sQ+P_1+\ldots+P_k$ for cyclic Γ

(1b) $D=rE+sV+tF+P_1+\ldots+P_k$ for non-cyclic Γ

where r, s, t are natural numbers and $P_1, \ldots, P_k$ are simple divisors of principal orbits. The preceding table 1 implies that D , given by (1a) or (1b), has trivial character if and only if the following divisibility conditions are fullfilled:

type of $G < SL(2,\mathbb{C})$	conditions for trivial character
cyclic of odd order n	$n \mid r-s$
cyclic of even order 2m	$2m \mid r-s+km$
binary q-dihedral, q odd	$4 \mid r-s+2(t+k)$
binary q-dihedral, q even	$2 \mid s+t$ and $2 \mid r+s$
binary tetrahedral	$3 \mid s-t$
binary octahedral	$2 \mid r+s$
binary icosahedral	—

Table 2

For every condition of table 2 we list finitely many (r,s, t,k) which generate the additive semigroup of all quadruples satisfying the condition. The corresponding divisors are listed in the following table 3:

type of G < SL(2,ℂ)	Invariant divisors with trivial character are sums of
cyclic of odd order n	nP, nQ, P+Q, P_i
cyclic of even order 2m	nP, nQ, P+Q, $mP+P_i$, $mQ+P_i$, P_i+P_j
binary q-dihedral, q odd	4E, 4V, E+V, 2E+F, 2V+F 2F, $2E+P_i$, $2V+P_i$, $F+P_i$, P_i+P_j
binary q-dihedral, q even	2E, 2V, 2F, E+V+F, P_i
binary tetrahedral	E, 3V, 3F, V+F, P_i
binary octahedral	2E, 2V, F, E+V, P_i
binary icosahedral	E, V, F, P_i

Table 3

Provisional Result: *By choosing one form for each divisor of the table we obtain a set of generators for the algebra of all* G-*invariant polynomials.*

§8 Generators and Relations for the Algebra of Invariant Polynomials

As in the preceding section $G < SU(2)$ will denote a finite subgroup whose image $\Gamma=\rho(G) < SO(3)$ acts on the projective line $\mathbb{P}=S^2$. If two semi-invariants have the same degree and the same character, any linear combination of them is again a semi-invariant having this degree and this character. In particular: All semi-invariants of degree $N=\#\Gamma$ and having the common character of principal divisors (§6, proposition 2) form a vector space W . A form f will be called *simple (principal)* if D_f is simple (principal). Thus {simple forms} $\subset$ {principal forms} $\subset$ W

Proposition 1: *The vector space* W *has dimension* 2 . *The simple forms span* W . *Every non-zero element of* W *is a principal form.*

Proof: Let V denote the sub vector space generated by the simple forms. Then $\dim V \geq 2$ because otherwise any two simple forms would be linearly dependent, i.e. there would be only one simple orbit, which is absurd. Let $[u,v] \subset W$ be the span of two linearly independent elements $u,v \in W$. Let Σ be a simple orbit.We choose a point $x \in \Sigma$ and find $\lambda,\mu \in \mathbb{C}$ such that $h=\lambda u+\mu v$ satisfies $h(x)=0$. Then $D_h=\Sigma$. This implies $V \subset [u,v]$, and so $V=[u,v]$ for dimension reasons. Thus $V=W$ and $\dim W=2$. Let a non-zero element $u \in W$ be given. There is a principal form v such that u and v have a common zero x . If u and v were linearly independent, every $w \in W=[u,v]$ would have a zero at this x , which is absurd. So $u=\lambda v$ for some $\lambda \in \mathbb{C}^{\times}$.

Let p, q, e, v, f denote forms whose divisors are P, Q, E, V, F, see the end of the preceding section. These forms are determined up to a factor in $\mathbb{C}^{\times}$. The algebra S^G of G-invariant polynomials in two variables is generated by the following invariant forms:

Type of $G < SL(2,\mathbb{C})$		Generators of S^G
C^n, cyclic of order n		p^n, q^n, pq
$\tilde{D}_q$, binary q-dihedral,	q odd	e^4, v^4, f^2, e^2f, v^2f, ev
	q even	e^2, v^2, f^2, evf
$\tilde{T}$, binary tetrahedral		e, v^3, f^3, vf
$\tilde{O}$, binary octahedral		e^2, v^2, f, ev
$\tilde{I}$, binary icosahedral		e, v, f

Table 1

Proof of table 1: The provisional result of the preceding section says that S^G is generated by the forms of table 1 together with further forms (like $p^m p_i$, $q^m p_i$, $p_i p_j$ for the cyclic group of order 2m) which have one or two factors p_i whose divisors are simple divisors fo principal orbits P_i . According to proposition 1 every p_i is a linear combination of the following two principal forms, which form a base of the vector space of all principal forms:

C_n, n odd	p^n q^n
C_{2m}	p^m, q^m
$\tilde{D}_q$	e^2, v^2
$\tilde{T}$, $\tilde{O}$, $\tilde{I}$	e^2, f^3

Table 2

Therefore all forms with a factor p_i can be dropped from the list of generators of S^G so that those listed in table 1 suffice.

Let $k=n(E)$, $m=n(V)$, and $n=n(F)$ denote the orders of the isotropy subgroups which belong to the orbits E, V, and F, see the tables at the end of I§6. Then e^k, v^m, and f^n are principal forms. They are linearly dependent, but any two of them are independent, thus $\alpha e^k+\beta v^m+\gamma f^n=0$ with coefficients $\alpha, \beta, \gamma \in \mathbb{C}^\times$. So far e, v, and f were only determined up to a factor in $\mathbb{C}^\times$. If these factors are suitably chosen the relation becomes $e^k+v^m+f^n=0$. We obtain:

$G < SL(2,\mathbb{C})$	Principal Relation
$\tilde{D}_q$	$e^2+v^2+f^q=0$
$\tilde{T}$	$e^2+v^3+f^3=0$
$\tilde{O}$	$e^2+v^4+f^3=0$
$\tilde{I}$	$e^2+v^5+f^3=0$

Table 3

Table 1 contains between three and six generators for S^G . The principal relations allow us to reduce the number of generators to three in each case. Here is a suitable choice:

		Generators of S^G	Relation $\varphi=0$
C_n		$x=p^n-q^n$, $y=p^n+q^n$, $z=pq$	$x^2-y^2+4z^n=0$
$\tilde{D}_q$,	q odd	$x=e^2f-v^2f$, $y=2ev$, $z=f^2$	$x^2+z(y^2-z^q)=0$
	q even	$x=2evf$, $y=e^2-v^2$, $z=f^2$	
$\tilde{T}$		$x=v^3-f^3$, $y=vf$, $z=e$	$x^2+4y^3-z^4=0$
$\tilde{O}$		$x=ev$, $y=f$, $z=e^2$	$x^2+z(z^2+y^3)=0$
$\tilde{I}$		$x=e$, $y=f$, $z=v$	$x^2+y^3+z^5=0$

Table 4

The three generators of table 4 determine a homomorphism $\Phi: \mathbb{C}[x,y,z] \to S$ = polynomial algebra in two variables. The image of Φ is the subalgebra S^G of G-invariant polynomials. The polynomial $\varphi(x,y,z)$ in the last column of table 4 is mapped to 0. Let $\langle\varphi\rangle \subset \mathbb{C}[x,y,z]$ denote the principal ideal generated by φ.

<u>Theorem</u>: *The homomorphism Φ induces an isomorphism*

$$\mathbb{C}[x,y,z]/\langle\varphi\rangle \cong S^G .$$

Proof: We use the notion of transcendence degree trd and some general results about it, see e.g. Zariski-Samuel I, Chap. II §12. The polynomials φ of table 4 are shown to be irreducible using that x occurs as term x^2 only. Then $\mathbb{C}[x,y,z]/\langle\varphi\rangle$ is entire, i.e. without zero divisors. As a subring of S the ring S^G is also entire, thus Φ induces an epimorphism

$$\bar{\Phi}: \mathbb{C}[x,y,z]/\langle\varphi\rangle \to S^G \tag{1}$$

between entire $\mathbb{C}$-algebras. We claim that

$$\operatorname{trd} S^G = 2 . \tag{2}$$

Proof: For any $u \in S$ the polynomial $h(X) = \prod_{\gamma\in G} (X-u\gamma) = X^N + a_{N-1}X^{N-1}+\ldots+a_o$ has coefficients $a_\nu \in S^G$. Since $h(u)=0$ this implies that S is an algebraic extension of S^G, in particular $\operatorname{trd} S^G = \operatorname{trd} S = 2$.

For an epimorphism $A \to B$ between entire rings $\operatorname{trd} A \geq \operatorname{trd} B$

with equality if and only if $A \to B$ is an isomorphism. Since the canonical epimorphism $\mathbb{C}[x,y,z] \to \mathbb{C}[x,y,z]/\langle\varphi\rangle$ is not an isomorphism $\operatorname{trd} \mathbb{C}[x,y,z]/\langle\varphi\rangle \leq 2$, on the other hand $\operatorname{trd} \mathbb{C}[x,y,z]/\langle\varphi\rangle \geq 2$ because of (1) and (2). Thus both $\mathbb{C}[x,y,z]/\langle\varphi\rangle$ and S^G have trd =2 and therefore $\bar{\Phi}$ is an isomorphism.

Appendix:

The principal relations of table 3 occur among the relations of table 4, if the group is changed and the forms are transformed as follows:

$\tilde{D}_q \to C_q$	$\tilde{T} \to \tilde{D}_2$	$\tilde{O} \to \tilde{T}$	$\tilde{I} \to \tilde{I}$
$e \to x$	$e \to \sqrt{2}x$	$e \to \frac{i}{2} x$	$e \to x$
$v \to iy$	$v \to \frac{i}{\sqrt{3}} y-x$	$v \to \frac{1}{\sqrt{2}} z$	$v \to z$
$f \to z$	$f \to -\frac{i}{\sqrt{3}}y-x$	$f \to -y$	$f \to y$

Table 5

This is no coincidence: Let $[G,G]$ denote the commutator subgroup of G so that $\mathcal{A}G = G/[G,G]$. Table 2 at the end of §5 implies the following table 6.

G	$\tilde{D}_q$	$\tilde{T}$	$\tilde{O}$	$\tilde{I}$
$[G,G]$	C_q	$\tilde{D}_2$	$\tilde{T}$	$\tilde{I}$

Table 6

This together with the following general result explains table 5.

Proposition: *Let* $G < GL(n,\mathbb{C})$ *be a finite subgroup. The algebra of* $[G,G]$*-invariant polynomials is generated by the* G-*semi-invariant forms.*

Proof: Every G-semi-invariant h is $[G,G]$-invariant, because the character χ_h factors through $\mathcal{A}G = G/[G,G]$. In order to prove that the G-semi-invariants generate all of $S^{[G,G]}$ it suffices to show that every $[G,G]$-invariant f is a sum of G-semi-invariants: Since f is $[G,G]$-invariant, $f\cdot\gamma$ is a well defined form for every $\gamma \in \mathcal{A}G$. Let $\chi\colon G \to \mathbb{C}^\times$

be a homomorphism. Then $f_\chi = \sum_{\gamma \in \mathcal{A}G} \chi(\gamma^{-1})(f\cdot\gamma)$ is G-semi-invariant with character χ. There are finitely many homomorphism $\chi_1,\ldots,\chi_s : \mathcal{A}G \to \mathbb{C}^\times$. Their sum satisfies

$$(*) \qquad \chi_1(\gamma)+\ldots+\chi_s(\gamma)=0 \quad \text{if} \quad \gamma \neq 1 .$$

Using the notion $f_j = f_{\chi_j}$ we obtain $f = \frac{1}{s}(f_1+\ldots+f_s)$.

In order to prove (*) let us first assume that $\mathcal{A}G$ is cyclic of order q, generated by α. Then χ is uniquely determined by $\chi(\alpha)$, and $\chi(\alpha)$ can be chosen to be any q-th root of unity. Hence there are q homomorphism determined by $\chi_j(\alpha) = \zeta^j$, $j=0,\ldots,q-1$, $\zeta = e^{2\pi i/q}$. For $\gamma=\alpha^k$ $(k=1,..,q-1)$ we obtain $\chi_o(\gamma)+\ldots+\chi_{q-1}(\gamma) = \chi_o(\alpha)^k+\ldots+\chi_{q-1}(\alpha)^k = 1+\zeta^k+\ldots+\zeta^{k(q-1)} = 0$, because the q-th roots of unity ζ^k $(k=1,..,q-1)$ are the roots of $1+z+\ldots+z^{q-1} = 0$. In general $\mathcal{A}G$ is a direct product of finite cyclic groups. Thus it suffices to show that (*) holdstrue for $\mathcal{A}G = A\times B$ assuming that it holds true for A and B. Let $\chi_1\ldots,\chi_s : A \to \mathbb{C}^\times$ and $\psi_1,\ldots,\psi_t : B \to \mathbb{C}^\times$ be all possible homomorphisms. Then the $\omega_{ij} : A\times B \to \mathbb{C}^\times$ defined by $\omega_{ij}(a,b) = \chi_i(a)\cdot\psi_j(b)$ are all possible homomorphisms of the direct product, if $i=1,\ldots s$ and $j=1,\ldots t$. We obtain $\sum_{i,j}\omega_{ij}(a,b) = \sum_i\chi_i(a)\cdot\sum_j\psi_j(b)$. Since $(a,b) \neq (1,1)$ at least one factor $=0$ by assumption.

§9 The Affine Orbit Variety

Behind our description of S^G as $\mathbb{C}[x,y,z]/\langle\varphi\rangle$, see the theorem in the preceding section, there is a general result for arbitrary finite subgroups of $GL(n,\mathbb{C})$.

<u>Proposition</u>: *For every finite subgroup* $G < GL(n,\mathbb{C})$ *the* $\mathbb{C}$-*algebra* S^G *of* G-*invariant polynomials in* n *variables is*

finitely generated.

The proof uses basic properties of noetherian rings, see e.g. Lang, Chap.VI. For $f \in S = \mathbb{C}[z_1,\dots,z_n]$ we consider the polynomial $\prod_{\gamma\in G}(X-f\gamma) = X^N + a_{N-1}X^{N-1} + \dots + a_o$ with coefficients $a_\nu \in S^G$. These coefficients depend on f, so we better write $a_\nu(f)$. Let $A \subset S^G$ denote the $\mathbb{C}$-subalgebra generated by $a_\nu(z_j)$, $j=1,\dots,n$ and $\nu=0,\dots,N-1$. Since A is finitely generated, it is noetherian. As A-module S is generated by finitely $z_1^{\nu_1}\dots z_n^{\nu_n}$ with $\nu_i < N$. Therefore the A-submodule S^G is also finitely generated. The generators of A as $\mathbb{C}$-algebra together with the generators of S^G as A-module generate S^G as $\mathbb{C}$-algebra.

We may assume that finitely many forms $1, f_1,\dots,f_r$ of degrees $0, d_1,\dots,d_r$ respectively generate the $\mathbb{C}$-algebra S^G. The polynomial map

(1) $$F: \mathbb{C}^n \to \mathbb{C}^r$$

with components $f_1,\dots,f_r$ induces the epimorphism

(2) $${}_rS = \mathbb{C}[x_1,\dots,x_r] \twoheadrightarrow S^G, \quad h \mapsto h\circ F,$$

and hence the isomorphism

(3) $${}_rS/\mathfrak{n} = S^G,$$

where $\mathfrak{n}$ = kernel of (2) is a finitely generated prime ideal. Thus the *affine orbit variety*

(4) $$V = \{x \in \mathbb{C}^r : h(x)=0 \text{ for every } h \in \mathfrak{n}\}$$

is irreducible. Further,

(5) $$\dim V = n .$$

For basic facts about affine varieties see e.g. Shafarevich, Chap.I. We have $\dim V = \operatorname{trd} {}_rS/\mathfrak{n} = \operatorname{trd} S^G = \operatorname{trd} {}_nS = n$ because ${}_nS$ is an algebraic extension of S^G.

The polynomial map F (1) has the following properties:

(6) $F(\mathbb{C}^n) = V$.

(7) $F(z)=F(w)$ if and only if $w=\gamma z$ for some $\gamma \in G$.

(8) F is proper (If K is compact, then $F^{-1}(K)$ is compact) and closed (If A is closed, then also $F(A)$)

(9) The map $F: \mathbb{C}^n \to V$ with restricted range is open.

Proof of (6): We have $F(\mathbb{C}^n) \subset V$ because $h \circ F = 0$ for every $h \in \mathfrak{a}$. In order to show $F(\mathbb{C}^n) \supset V$ let $c = (c_1, \ldots, c_r) \in V$ be given. Let $\mathfrak{b} \in {}_nS$ denote the ideal generated by $f_1 - c_1, \ldots, f_r - c_r$. If $\mathfrak{b} \neq {}_nS$ there is a common zero $b \in \mathbb{C}^n$ of $f_1 - c_1, \ldots, f_r - c_r$ by Hilbert's Nullstellensatz, hence $F(b) = c$. It remains to show that $\mathfrak{b} = {}_nS$ is impossible: The assumption $\mathfrak{b} = {}_nS$ implies that there are $p_1, \ldots, p_r \in {}_nS$ with $p_1 \cdot (f_1 - c_1) + \ldots + p_r \cdot (f_r - c_r) = 1$. We transform with the elements $\gamma \in G$ and add, using the invariance of f_i:

$$\left(\sum_{\gamma \in G} p_1 \gamma\right)(f_1 - c_1) + \ldots + \left(\sum_{\gamma \in G} p_r \gamma\right)(f_r - c_r) = \# G$$

Since $\Sigma p_i \gamma \in S^G$ there are $\tilde{p}_i \in {}_rS$ with $\tilde{p}_i \cdot F = \Sigma p_i \gamma$, thus $(\tilde{p}_1 \cdot (x_1 - c_1) + \ldots + \tilde{p}_r \cdot (x_r - c_r)) \circ F = \# G$, i.e. $h(x) = \tilde{p}_1 \cdot (x_1 - c_1) + \ldots + p_r \cdot (x_r - c_r) - \# G \in \mathfrak{a}$. Therefore $h(x) = 0$ for every $x \in V$, in particular $h(c) = 0$. This implies $\# G = 0$, a contradiction.

Proof of (7): If $w = \gamma z$, then $F(w) = F(z)$ because the components f_i of F are G-invariant. Assume now $w \neq \gamma z$ for every $\gamma \in G$. There is an $f \in {}_nS$ with $f(z) = 0$ and $f(\gamma w) = 1$ for every $\gamma \in G$. Since $\prod_{\gamma \in G} f\gamma \in S^G$, there is an $h \in {}_rS$ with $h \circ F = \Pi f \gamma$. Then $h \circ F(w) = 1$, and $h \circ F(z) = 0$ whence $F(w) \neq F(z)$.

Proof of (8): For $t \in \mathbb{C}$ and $y = (y_1, \ldots, y_r) \in \mathbb{C}^r$ let

$$s(t,y) = (t^{d_1} y_1, \ldots, t^{d_r} y_r) \tag{10}$$

Then $F(tx) = s(t, F(x))$. Let $d = \operatorname{lcm}(d_1, \ldots, d_r)$ and $a_i = d/d_i$. The following quasi-norm on $\mathbb{C}^r$ will be used:

$$\sigma(y) = \sqrt[d]{|y_1|^{a_1} + \ldots + |y_r|^{a_r}} \tag{11}$$

For every neighbourhood U of 0 in $\mathbb{C}^r$ there is an $\varepsilon > 0$ such that $\{y \in \mathbb{C}^r : \sigma(y) < \varepsilon\} \subset U$. We have $\sigma(s(t,y)) = |t| \sigma(y)$ and so

$$\sigma(F(tx)) = |t| \sigma(F(x)) \quad \text{for } x \in \mathbb{C}^n \text{ and } t \in \mathbb{C}. \tag{12}$$

Let $\|x\| = \sqrt{|x_1|^2 + \ldots + |x_n|^2}$ denote the usual Hermitean norm on $\mathbb{C}^n$, let $S^{2n-1} = \{x \in \mathbb{C}^n : \|x\| = 1\}$ denote the corresponding compact unit sphere. Its image $F(S^{2n-1}) \subset \mathbb{C}^r$ is compact and does not contain the origin because $F(x)=0$ if and only if $x=0$, see (7). Hence there is a $\delta>0$ such that $\sigma(F(x))>\delta$ for all $x \in S^{2n-1}$. Together with (12) this yields

$$\sigma(F(x))>\delta\|x\| \quad \text{for all} \quad x\in\mathbb{C}^n . \tag{13}$$

Let $B\subset\mathbb{C}^r$ be compact, i.e. closed and bounded. Then $F^{-1}(B)$ is closed because F is continuous, and $F^{-1}(B)$ is bounded because of (13).
Let $A\subset\mathbb{C}^n$ be closed and let $c\in\mathbb{C}^r\setminus F(A)$. In order to find a neighbourhood U of c such that $U\cap F(A)=\emptyset$ we begin with the (big) neighbourhood $W=\{y\in\mathbb{C}^r:\ \sigma(y)<2\|c\|\}$. The compact set $K=\{x\in\mathbb{C}^n:\ \|x\|<2\|c\|/\delta\}$ contains $F^{-1}(W)$ because of (13). The image $F(A\cap K)$ is compact whence closed: There is a neighbourhood W' of c such that $W'\cap F(A\cap K)=0$. Then $U=W\cap W'$ does not meet $F(A)$. Property (9) follows from (6)-(8). Together they imply

Theorem: *The polynomial map* F, *defined by* (1), *induces an homeomorphism* $\bar{F}:\ \mathbb{C}^n/G \approx V$ *between the orbit space* $\mathbb{C}^n/G$ *and the affine variety* V. Therefore V is called the *affine orbit variety.*

The affine orbit varieties of conjugate subgroups of $GL(n,\mathbb{C})$ *are isomorphic* : Let $h\in GL(n,\mathbb{C})$ and let $G'=hGh^{-1}$. Let $F':\ \mathbb{C}^n\to\mathbb{C}^s$ be a G'-invariant polynomial map which determines the orbit variety $V'=F'(\mathbb{C}^n)$ of the G'-action. There are polynomial maps such that the following diagram is commutative:

$$\begin{array}{ccccc}
\mathbb{C}^n & \xrightarrow{\ h\ } & \mathbb{C}^n & \xleftarrow{\ h\ } & \mathbb{C}^n \\
{\scriptstyle F}\downarrow & & \downarrow{\scriptstyle F'} & & \downarrow{\scriptstyle F} \\
\mathbb{C}^r & \xrightarrow{\ \Phi\ } & \mathbb{C}^s & \xrightarrow{\ \Psi\ } & \mathbb{C}^r
\end{array}$$

The components φ_i and ψ_j of Φ and Ψ are obtained as follows: For every component f_i' of F' the composition $f_i'\circ h$ is G-invariant, hence $f_i'\circ h=\varphi_i\circ F$ for some $\varphi_i\in {}_rS$. For every component f_j of F the composition $f_j\circ h^{-1}$ is G'-invariant, hence $f_j\circ h^{-1}=\psi_j\circ F'$ for some $\psi_j\in {}_sS$. The diagram implies that Φ maps V isomorphically onto V' with inverse $(\Phi|V)^{-1}=\Psi|V'$.

CHAPTER III

LOCAL THEORY OF SEVERAL COMPLEX VARIABLES

We want to study the quotient varieties $\mathbb{C}^2/G$ of the finite subgroups of $SL(2,\mathbb{C})$ and their singularities from the viewpoint of complex analytic geometry. Further, we like to show how they fit into the general context of singularities. This requires some knowledge of complex analysis of several variables. This chapter does not present a full discussion of the topic. It is rather a concise introduction which covers the general background as well as special results for later use. Since 'very local' results will suffice, no coherent sheaves will occur.

There is no lack of books on several complex variables. We shall refer to them for the proofs of the basic facts of the theory. Essentially all results of this chapter including their proofs can be found in the literature, but often not in the form which we want for later use. Therefore the local structure of finite maps is presented in some detail including the necessary proofs. The reader who knows the subject may of course skip this chapter and may look up only special results when they are employed later.

§1 Germs of Holomorphic Functions

Let X and Y be topological spaces, let $a \in X$. In order to define the *germ of a map* $\varphi: (X,a) \to Y$ we consider all open neighbourhoods U of a in X and all maps $f: U \to Y$.

Two such maps $f_1: U_1 \to Y$ and $f_2: U_2 \to Y$ are equivalent if there is a smaller open neighbourhood U such that $a \in U \subset \subset U_1 \cap U_2$ and $f_1|U = f_2|U$. The equivalence classes are called germs of maps $(X,a) \to Y$. The value $\varphi(a) \in Y$ is well defined. Mostly a map $U \to Y$ and the germ represented by it are denoted by the same symbol.

In this section we consider only $X=\mathbb{C}^n$, $Y=\mathbb{C}$ and holomorphic functions $U \to \mathbb{C}$. They form a $\mathbb{C}$-algebra $\mathcal{O}(U)$. The *germs of holomorphic functions* $(\mathbb{C}^n,a) \to \mathbb{C}$ form also a $\mathbb{C}$-algebra which is denoted by ${}_n\mathcal{O}_a$.

For simplicity we restrict to the point $a=0$ and write ${}_n\mathcal{O} = {}_n\mathcal{O}_o$. The holomorphic germs $f \in {}_n\mathcal{O}$ can be identified with the convergent power series in n variables $z=(z_1,..,z_n)$,

$$\sum_{\nu \in \mathbb{N}^n} a_\nu z^\nu \quad , \quad a_\nu \in \mathbb{C} \, .$$

Here $\nu=(\nu_1,..,\nu_n) \in \mathbb{N}^n$ is a multiindex, and $z^\nu = z_1^{\nu_1}..z_n^{\nu_n}$. The polynomial algebra $\mathbb{C}[z_1,..,z_n]$ is a subalgebra of ${}_n\mathcal{O}$.

The $f \in {}_n\mathcal{O}$ with $f(0) \neq 0$ are units, the $f \in {}_n\mathcal{O}$ with $f(0)=0$ form the unique maximal ideal ${}_n\mathfrak{m} \subset {}_n\mathcal{O}$. Thus ${}_n\mathcal{O}$ is a local algebra. The maximal ideal ${}_n\mathfrak{m}$ is generated by the n coordinate functions $z_1,\ldots,z_n$. The kth power ${}_n\mathfrak{m}^k$ is generated by all monomials z^ν with $|\nu| = \nu_1+\ldots+\nu_n = k$. An $f \in {}_n\mathcal{O}$ belongs to ${}_n\mathfrak{m}^k$ if and only if its Taylor series begins with terms of order k (If $f = \Sigma a_\nu z^\nu$ then $a_\nu = 0$ for $|\nu|<k$) . The powers form a decreasing sequence

$${}_n\mathcal{O} \supset {}_n\mathfrak{m} \supset {}_n\mathfrak{m}^2 \supset \ \ldots \ \text{with} \ \bigcap_{k=1}^{\infty} {}_n\mathfrak{m}^k = 0 \ .$$

The elements of the factor algebra ${}_n\mathcal{O}/{}_n\mathcal{O}^{k+1}$ can be identified with the polynomials of degree $\leq k$. Hence ${}_n\mathcal{O}/{}_n\mathfrak{m}^{k+1}$ is finite dimensional $\mathbb{C}$-algebra. There are isomorphisms

$$_n\mathcal{O}/_n\mathfrak{m} \cong \mathbb{C} \quad \text{induced by} \quad f \mapsto f(0) \quad \text{and}$$
$$_n\mathfrak{m}/_n\mathfrak{m}^2 \cong \mathbb{C}^n \quad \text{induced by} \quad f \mapsto \left(\frac{\partial f}{\partial z_1}(0),\ldots,\frac{\partial f}{\partial z_n}(0)\right).$$

Identity Theorem: *Let* $U \subset \mathbb{C}^n$ *be open and connected, let* $f,g \in \mathcal{O}(U)$. *If* $f|V=g|V$ *for some open non-empty subset* U *then* $f=g$.

Corollary 1: *Let* $a \in U$. *The homomorphism* $\mathcal{O}(U) \to {}_n\mathcal{O}_a$, $f \mapsto$ germ of f at a , *is injective.*

Corollary 2: *The algebra* $_n\mathcal{O}$ *is entire.*

We come to the basic theorems about $_n\mathcal{O}$. In order to distinguish the last coordinate of $\mathbb{C}^n$ we denote the coordinates by $(z_1,\ldots,z_{n-1},t)$. The germs of functions which do not depend on t form a subalgebra $_{n-1}\mathcal{O} = \mathcal{O}' \subset {}_n\mathcal{O}$. The germ $f \in {}_n\mathcal{O}$ is called *regular with respect to* t if $f(0,\ldots,0,t) \neq 0$. Every germ $f \neq 0$ becomes regular with respect to t after a linear change of coordinates. If f is regular, we have $f(0,\ldots 0,t) = at^r$ + higher order terms with $0 \neq a \in \mathbb{C}$.

Division Theorem: *Let* $f \in {}_n\mathcal{O}$ *be regular of order* r *with respect to* t . *Then every* $h \in {}_n\mathcal{O}$ *can be written as*

(1) $$h = g \cdot f + a_{r-1}t^{r-1} + \ldots + a_o$$

where $g \in {}_n\mathcal{O}$ *and* $a_o,\ldots,a_{r-1} \in \mathcal{O}'$ *are uniquely determined by* h *and* f .

There are essentially two proofs available, either by power series, see e.g. Hörmander or Grauert and Remmert or by Cauchy's Integral Formula, see e.g. Gunning and Rossi.

The remainder $a_{r-1}t^{r-1} + \ldots + a_o \in \mathcal{O}'[t]$ is a polynomial in t of order $<r$ with coefficients in $\mathcal{O}' = {}_{n-1}\mathcal{O}$. If for a polynomial $p \in \mathcal{O}'[t]$ the highest coefficient is $=1$ (p is "monic"), and if all the other coefficients belong to the maximal ideal $_{n-1}\mathfrak{m} \subset {}_{n-1}\mathcal{O} = \mathcal{O}'$, then p is called a *Weierstrass*

polynomial. They occur in Weierstrass'

Vorbereitungssatz: *If* $f \in {}_n\mathcal{O}$ *is regular of order* r *with respect to* t *then*

$$f = u \cdot p \quad ,$$

where $u \in {}_n\mathcal{O}$ *is a unit and* p *is a Weierstrass polynomial of order* r . *Both* u *and* p *are uniquely determined by* f .

The Vorbereitungssatz is an easy consequence of the Division Theorem applied to $h=t^r$. The Division Theorem and the Vorbereitungssatz are used in order to prove by induction on n :

Theorem: *The* $\mathbb{C}$*-algebra* ${}_n\mathcal{O}$ *is factorial and noetherian.* "Factorial" means that unique factorization into irreducible elements holds true. "Noetherian" means that every submodule of a finitely generated ${}_n\mathcal{O}$-module is also finitely generated; particularly, every ideal in ${}_n\mathcal{O}$ is finitely generated. Every textbook on Several Complex Variables contains a proof of this theorem.

In the following sections the ideals $\mathfrak{a} \subset {}_n\mathcal{O}$ and their *residue class rings* ${}_n\mathcal{O}/\mathfrak{a}$ are important. As ${}_n\mathcal{O}$ the ${}_n\mathcal{O}/\mathfrak{a}$ are $\mathbb{C}$-algebras, they are noetherian and local. The maximal ideal of ${}_n\mathcal{O}/\mathfrak{a}$ is $\mathfrak{m}' = {}_n\mathfrak{m}/\mathfrak{a}$ (except for the trivial case $\mathfrak{a} = {}_n\mathcal{O}$) . The powers are $\mathfrak{m}'^k = (\mathfrak{m}^k + \mathfrak{a})/\mathfrak{a}$. They form a decreasing sequence ${}_n\mathcal{O}/\mathfrak{a} \supset \mathfrak{m}' \supset \mathfrak{m}'^2 \supset \ldots \supset \mathfrak{m}'^k \supset \ldots$ with $\bigcap_{k=1}^{\infty} \mathfrak{m}'^k = 0$.

§2 Germs of Analytic Sets

Two subsets X and Y of $\mathbb{C}^n$ are said to be equivalent at 0 if there is a neighbourhood U of 0 such that $X \cap U = Y \cap U$. An equivalence class of subsets is called *the germ* (at 0) *of a set*. Usually subsets and their germs are denoted by the same

§3 Germs of Holomorphic Maps

The *germ* $F: (\mathbb{C}^n,a) \to (\mathbb{C}^q,b)$ *of a map* is uniquely determined by its components $f_1,..,f_q: (\mathbb{C}^n,a) \to (\mathbb{C},b)$, which are germs of functions with $f_i(a)=b$. The germ F is called *holomorphic* if every f_i belongs to ${}_n\mathcal{O}_a$. The *differential* of F at a is the complex linear map $D_aF: \mathbb{C}^n \to \mathbb{C}^q$ which is given by the *Jacobian* (n×q)-matrix of partial derivatives $\partial f_i/\partial z_k$ at a where $i=1,..,q$ and $k=1,..,n$. We shall often use the following facts whose real analogs are well known. For the complex case see e.g. the book by Gunning and Rossi, Chap.I Sec.B.

<u>Chain Rule</u>: *A holomorphic germ* F *as above can be composed with a holomorphic germ* $G: (\mathbb{C}^q,b) \to (\mathbb{C}^r,c)$ *to a holomorphic germ* $G\circ F: (\mathbb{C}^n,a) \to (\mathbb{C}^r,c)$. *The differential of the composition is the composition of the differentials,*

$$D_a(G\circ F) = D_bG\circ D_aF \quad .$$

<u>Inverse Mapping Theorem</u>: *If* $q=n$ *and* D_aF *is invertible, then* F *is invertible: There is a holomorphic germ* $G: (\mathbb{C}^n,b) \to (\mathbb{C}^n,a)$ *such that* $G\circ F = id$ *and* $F\circ G = id$.

In this case F is called *biholomorphic*. If in addition $b=0$, the components $f_1,..,f_n$ of F are called *local coordinates* of $\mathbb{C}^n$ at a .

<u>Submersions</u>: A holomorphic germ $F: (\mathbb{C}^n,a) \to (\mathbb{C}^q,b)$ is called a *submersion* if $q\le n$ and the rank $\operatorname{rk} D_aF = q$. If the germs $f_1,..,f_q \in {}_n\mathfrak{m}_a$ are the components of a submersion, they can be completed to a local coordinate system $(f_1,..,f_q,f_{q+1},..f_n)$ of $\mathbb{C}^n$ at a . A submersion F is locally open: For a representative $F: U \to \mathbb{C}^q$ the image $F(U)$ is a neighbourhood of b.

<u>Immersions</u>: A holomorphic germ $F: (\mathbb{C}^n,a) \to (\mathbb{C}^q,b)$ is called an *immersion* if $n\le q$ and $\operatorname{rk} D_aF = n$. In this case there is a submersion $G: (\mathbb{C}^q,b) \to (\mathbb{C}^n,a)$ such that $G\circ F = id$.

The holomorphic germ $F: (\mathbb{C}^n,a) \to (\mathbb{C}^q,b)$ induces a homomorphism of $\mathbb{C}$-algebras

$$F^a: {}_q\mathcal{O}_b \to {}_n\mathcal{O}_a \,, \quad F^a(h) = h\circ F \;,$$

which is local, i.e. $F^a({}_q\mathfrak{m}_b) \subset {}_n\mathfrak{m}_a$. For simplicity we restrict from now on to the case a=0, b=0, and omit the lower indices a and b. The differential DF is determined by F^a as follows: In the commutative diagram below the vertical isomorphisms are induced by ${}_n\mathfrak{m} \to \mathbb{C}^n$, $f \mapsto \operatorname{grad} f(0)$. The upper $\mathbb{C}$-linear map F^* is induced by F^a, the lower map $(DF)^\vee$ is the dual of DF.

$$\begin{array}{ccc} {}_q\mathfrak{m}/{}_q\mathfrak{m}^2 & \xrightarrow{F^*} & {}_n\mathfrak{m}/{}_n\mathfrak{m}^2 \\ \wr\Vert & & \wr\Vert \\ \mathbb{C}^q & \xrightarrow{(DF)^\vee} & \mathbb{C}^n \end{array} .$$

In particular, F is biholomorphic if and only if F^* is an isomorphism.

<u>Theorem</u>: *Let $\mathfrak{a} \subset {}_n\mathcal{O}$ and $\mathfrak{b} \subset {}_q\mathcal{O}$ be two ideals. Every homomorphism of $\mathbb{C}$-algebras $\Phi: {}_q\mathcal{O}/\mathfrak{b} \to {}_n\mathcal{O}/\mathfrak{a}$ is induced by a holomorphic germ $F: (\mathbb{C}^n,0) \to (\mathbb{C}^q,0)$.*

Proof: We may assume $\mathfrak{a} \subset {}_n\mathfrak{m}$. We may further assume $\mathfrak{b} = \langle 0\rangle$ because an F, which induces ${}_q\mathcal{O} \twoheadrightarrow {}_q\mathcal{O}/\mathfrak{b} \xrightarrow{\Phi} {}_n\mathcal{O}/\mathfrak{a}$, also induces Φ. Let $y_1,\ldots,y_q$ denote the coordinate functions of $\mathbb{C}^q$. For every $\Phi(y_i)$ we choose an $f_i \in {}_n\mathcal{O}$ such that $\Phi(y_i)$ is the image of f_i in ${}_n\mathcal{O}/\mathfrak{a}$. The mapping germ F with components $f_1,\ldots,f_q$ has the desired property: The $\mathbb{C}$-algebra map $F^*: {}_q\mathcal{O} \xrightarrow{F^o} {}_n\mathcal{O} \twoheadrightarrow {}_n\mathcal{O}/\mathfrak{a}$ coincides with Φ on the subalgebra of polynomials $\mathbb{C}[y_1,\ldots,y_q] \subset {}_q\mathcal{O}$. In order to extend this to all of ${}_q\mathcal{O}$ we first observe that not only F^* but also Φ is local, i.e. $\Phi({}_q\mathfrak{m}) \subset \mathfrak{m}' := {}_n\mathfrak{m}/\mathfrak{a}$: Let $f \in {}_q\mathfrak{m}$. Then $\Phi(f) = \gamma + g$ with $\gamma \in \mathbb{C}$ and $g \in \mathfrak{m}'$. We consider $\Phi(-\gamma+f) = g$. If $\gamma \neq 0$, then $-\gamma+f$ is a unit and so g would be a unit. Therefore $\gamma = 0$. We shall use the following consequence of this result:

$$F^*({}_q\mathfrak{m}^k) \subset \mathfrak{m}'^k \quad \text{and} \quad \Phi({}_q\mathfrak{m}^k) \subset \mathfrak{m}'^k \quad \text{for every } k .$$

The Taylor expansion yields for every $f \in {}_q\mathcal{O}$ a sequence of polynomials $f_k \in \mathbb{C}[y_1,\ldots,y_q]$ such that $f - f_k \in {}_q\mathfrak{m}^k$. Therefore $F^*(f) - \Phi(f) = F^*(f-f_k) - \Phi(f-f_k) \in \mathfrak{m}'^k$ for every k and so $F^*(f) - \Phi(f) \in \bigcap_{k=1}^{\infty} \mathfrak{m}'^k = \langle 0\rangle$.

Let X and Y be analytic sets in $\mathbb{C}^n$ and $\mathbb{C}^q$ respectively which contain the origins. The germ of a map $(X,0) \to (Y,0)$ depends only on the germs of X and Y at 0.

For a holomorphic germ $G: (\mathbb{C}^n,0) \to (\mathbb{C}^q,0)$ the restriction $G|X: (X,0) \to (\mathbb{C}^q,0)$ is defined in the same way as for functions, see §2. The range of $G|X$ is $(Y,0)$ if and only if $G^o(\mathfrak{I}(Y)) = \langle 0 \rangle$.

The germ $F: (X,0) \to (Y,0)$ is called *holomorphic* if $F = G|X$ for some holomorphic germ $G: (\mathbb{C}^n,0) \to (\mathbb{C}^q,0)$. Every holomorphic F induces the $\mathbb{C}$-algebra homomorphism $F^o: \mathcal{O}_{Y,0} \to \mathcal{O}_{X,0}$ $h \mapsto h \circ F$. The germ F is uniquely determined by F^o. The theorem applied to $\mathfrak{a} = \mathfrak{I}(X)$ and $\mathfrak{b} = \mathfrak{I}(Y)$ shows that every homomorphism of $\mathbb{C}$-algebras $\Phi: \mathcal{O}_{Y,0} \to \mathcal{O}_{X,0}$ is induced by a holomorphic germ $F: (X,0) \to (Y,0)$, i.e. $\Phi = F^o$.

The germs $(X,0) \subset (\mathbb{C}^n,0)$ and $(Y,0) \subset (\mathbb{C}^q,0)$ are said to be *isomorphic* if there are holomorphic germs $F: (X,0) \to (Y,0)$ and $G: (Y,0) \to (X,0)$ with $G \circ F = \mathrm{id}_X$ and $F \circ G = \mathrm{id}_Y$. In this case the germ F is called *biholomorphic*. The theorem implies the

Corollary: *The germs $(X,0)$ and $(Y,0)$ are isomorphic if and only if the $\mathbb{C}$-algebras $\mathcal{O}_{X,0}$ and $\mathcal{O}_{Y,0}$ are isomorphic.*

§4 The Embedding Dimension

Let r holomorphic germs $f_1,\dots,f_r \in {}_n\mathfrak{m}$ be given. We consider the mapping germ $F: (\mathbb{C}^n,0) \to (\mathbb{C}^r,0)$ with components f_i and the ideal $\mathfrak{a} = \langle f_1,\dots,f_r\rangle$ generated by $f_1,\dots,f_r$. The *rank* $\operatorname{rk} F$ is defined to be the rank of the differential D_oF. The quotient algebra $\mathcal{O}' = {}_n\mathcal{O}/\mathfrak{a}$ is local with maximal ideal $\mathfrak{m}' = {}_n\mathfrak{m}/\mathfrak{a}$. The following numbers are equal:

(1) $$n - \operatorname{rk} F = \dim_{\mathbb{C}} {}_n\mathfrak{m}/(\mathfrak{a} + {}_n\mathfrak{m}^2) = \dim_{\mathbb{C}} \mathfrak{m}'/\mathfrak{m}'^2 .$$

This number is called the *corank* $\operatorname{crk} F$ of F or the *embedding dimension* $\operatorname{edim}\mathfrak{a}$ of $\mathfrak{a}$. The second equality follows because the projection ${}_n\mathfrak{m} \to \mathfrak{m}'$ induces an isomorphism ${}_n\mathfrak{m}/(\mathfrak{a} + {}_n\mathfrak{m}^2) \cong \mathfrak{m}'/\mathfrak{m}'^2$. In order to prove the first equality we remember (§3) that $\operatorname{rk} F$ is the rank of the linear map $F^*: {}_r\mathfrak{m}/{}_r\mathfrak{m}^2 \to {}_n\mathfrak{m}/{}_n\mathfrak{m}^2$ induced by F. The image of F^* is the first term in the exact sequence

$$0 \to (\mathfrak{a} + {}_n\mathfrak{m}^2)/{}_n\mathfrak{m}^2 \to {}_n\mathfrak{m}/{}_n\mathfrak{m}^2 \to {}_n\mathfrak{m}/(\mathfrak{a} + {}_n\mathfrak{m}^2) \to 0$$

Obviously $\operatorname{edim}\mathfrak{a} \le n$, and $= n$ if and only if $\mathfrak{a} \subset {}_n\mathfrak{m}^2$. In this case $\mathfrak{a}$ is called *minimally embedded*.

Let $\operatorname{edim}\mathfrak{a} = k$ for the ideal $\mathfrak{a} = \langle f_1,\dots,f_r\rangle \subset {}_n\mathfrak{m}$, let $q = n-k$. We may assume that the differential of the first q generators $f_1,\dots,f_q$ has rank q. A new local coordinate system of $(\mathbb{C}^n,0)$ can be chosen such that $z_1=f_1,\dots,z_q=f_q$ are the first q coordinates. Then $V = N(f_1,\dots,f_q) = \mathbb{C}^k$ is the subspace of the last k coordinates. There is an epimorphism

(2) $${}_n\mathcal{O} \twoheadrightarrow {}_k\mathcal{O}, \quad \varphi \mapsto \varphi|V .$$

The image $\mathfrak{a}' \subset {}_k\mathcal{O}$ of $\mathfrak{a}$ is generated by $f_{q+1}|V,\dots,f_r|V$. Vice versa, if $\mathfrak{a}'$ is generated by $g_1',\dots,g_s'$, there are elements $g_1,\dots g_s \in \mathfrak{a}$ with $g_i|V=g_i'$, and $\mathfrak{a}$ is generated by $f_1,\dots,f_q, g_1,\dots,g_s$. We have

(3) $$N(\mathfrak{a}) = N(\mathfrak{a}') \subset V = \mathbb{C}^k .$$

The epimorphism (2) induces an isomorphism

(4) $${}_n\mathcal{O}/\mathfrak{a} \cong {}_k\mathcal{O}/\mathfrak{a}' .$$

The partial derivatives $\partial f_i/\partial z_j$ at 0 vanish for $i,j>q$. Therefore $f_i|V \subset {}_k\mathfrak{m}^2$ for $i>q$ and hence $\mathfrak{a}'$ is minimally embedded.

Proposition: *Assume that* ${}_n\mathcal{O}/\mathfrak{a} \cong {}_m\mathcal{O}/\mathfrak{b}$ *for two minimally embedded ideals* $\mathfrak{a}$ *and* $\mathfrak{b}$. *Then* $m=n$, *and the isomorphism is induced by a biholomorphic germ* $F: (\mathbb{C}^n,0) \to (\mathbb{C}^n,0)$, *whose induced automorphism* $F^o: {}_n\mathcal{O} \to {}_n\mathcal{O}$ *maps* $\mathfrak{b}$ *isomorphically onto* $\mathfrak{a}$.

Proof: We have $n = \operatorname{edim}\mathfrak{a} = \operatorname{edim}\mathfrak{b} = m$. Let F be the germ of a map, such that F^o induces the isomorphism ${}_n\mathcal{O}/\mathfrak{b} \cong {}_n\mathcal{O}/\mathfrak{a}$, see §3, theorem 1. Then F^o induces also an isomorphism between ${}_n\mathfrak{m}/({}_n\mathfrak{m}^2+\mathfrak{b})$ and ${}_n\mathfrak{m}/({}_n\mathfrak{m}^2+\mathfrak{a})$. Since $\mathfrak{a}$ and $\mathfrak{b}$ are contained in ${}_n\mathfrak{m}^2$, being minimally embedded, this is an automorphism of ${}_n\mathfrak{m}/{}_n\mathfrak{m}^2$. Therefore F is biholomorphic by the Inverse Mapping Theorem, see the remark before the theorem in §3. The induced automorphism F^o of ${}_n\mathcal{O}$ maps $\mathfrak{b}$ isomorphically onto $\mathfrak{a}$.

The *embedding dimension* of the germ $(X,0) \subset (\mathbb{C}^n,0)$ of an analytic set is defined to be the embedding dimension of its ideal, viz.

$$\operatorname{edim} X = \operatorname{edim} \mathfrak{J}(X) .$$

The germ X is called minimally embedded if $\mathfrak{J}(X)$ is minimally embedded. If $\operatorname{edim} X = k$, there is a local coordinate system of $(\mathbb{C}^n,0)$ such that X is contained in the subspace $\mathbb{C}^k$ of the last k coordinates, and $(X,0) \subset (\mathbb{C}^k,0)$ is minimally embedded. In general, a minimally embedding $(X,0) \subset (\mathbb{C}^n,0)$ implies $m \geq n$ for every analytic germ $(Y,0) \subset (\mathbb{C}^m,0)$ which is isomorphic to X because then $\operatorname{edim} X = \operatorname{edim} Y$. Shortly we may say:

The embedding dimension of an analytic germ X *is the smallest possible* k *such that* X *can be embedded in* $\mathbb{C}^k$.

<u>Warning</u>: The embedding dimension of an ideal $\mathfrak{a}$ and the embedding dimension of its zero set $N(\mathfrak{a})$ may differ, consider e.g. $\mathfrak{a} = {}_n\mathfrak{m}^2$.

An ideal $\mathfrak{a} \subset {}_n\mathcal{O}$ is called *regular* if ${}_n\mathcal{O}/\mathfrak{a} \cong {}_k\mathcal{O}$ for some k . Equivalent statements are:

(5) $\mathfrak{a} = \langle z_1,\dots,z_q\rangle$ for some local coordinate system $(z_1,\dots,z_n)$ of $(\mathbb{C}^n,0)$.

(6) $\operatorname{edim} \mathfrak{a} = k$ and $\mathfrak{a}$ is generated by $n-k$ elements.

Regular ideals are prime .

An analytic germ $(X,0)$ is called *regular* if it is isomorphic to $(\mathbb{C}^k,0)$ for some k . This is the case if and only if $\mathcal{J}(X)$ is regular.

A minimally embedded regular ideal is zero . A minimally embedded regular germ $(X,0) \subset (\mathbb{C}^k,0)$ equals $(\mathbb{C}^k,0)$.

If $f_1,\dots,f_q \in {}_n\mathfrak{m}$ are the components of a submersion, $N(f_1,\dots,f_q)$ is isomorphic to $(\mathbb{C}^{n-q},0)$.
An immersion $(\mathbb{C}^r,0) \to (\mathbb{C}^n,0)$ maps $(\mathbb{C}^r,0)$ isomorphically onto the germ of an analytic subset.

§5 The Preparation Theorem

An important tool in order to study modules M over a local ring R with maximal ideal $\mathfrak{m}$ is

<u>Nakayama's Lemma</u>: *If* M *is finitely generated and* $M=N+\mathfrak{m}M$ *for a submodule* N , *then* $M=N$.

Proof: The quotient module $Q=M/N$ is also finitely generated, and $Q=\mathfrak{m}Q$. We show that $Q=0$. Assume that $Q\neq 0$. Let $\{u_1,\ldots,u_r\}$ be a minimal set of generators of Q . Since $Q=\mathfrak{m}Q$ there is an equation $u_r=\alpha_1u_1+\ldots+\alpha_ru_r$ with $\alpha_i\in\mathfrak{m}$. This can be written as $(1-\alpha_r)u_r=\alpha_1u_1+\ldots+\alpha_{r-1}u_{r-1}$. Since $1-\alpha_r$ is a unit, the elements $u_1,\ldots,u_{r-1}$ would already generate Q .

The quotient module $M/\mathfrak{m}M$ is a vector space over the residue class field $k=R/\mathfrak{m}$. We have $M/\mathfrak{m}M = M\otimes_R k$.

<u>Corollary</u>: *Let* M *be finitely generated over* R . *The elements* $u_1,\ldots,u_r$ *generate* M *as* R*-module if and only if their residue classes generate* $M/\mathfrak{m}M$ *as vector space over* k . *In particular: The minimal number of generators is* $\dim_k M/\mathfrak{m}M$.

Proof: One direction (if $u_1,\ldots,u_r$ generate) is obvious. Let the residue classes of $u_1,\ldots,u_r\in M$ generate $M/\mathfrak{m}M$. Let $N\subset M$ be the submodule generated by $u_1,\ldots,u_r$. Then $M = N+\mathfrak{m}M$, and so $M=N$ by the lemma.

Let $F: (\mathbb{C}^n,0) \to (\mathbb{C}^q,0)$ be a holomorphic germ and let M be an ${}_n\mathcal{O}$-module. Then M is also a module over the subalgebra $R=F^\circ({}_q\mathcal{O}) \subset {}_n\mathcal{O}$ and therefore a module over ${}_q\mathcal{O}$. We consider the ${}_q\mathcal{O}$-submodule ${}_q\mathfrak{m}\cdot M$. It is also an ${}_n\mathcal{O}$-submodule, namely ${}_q\mathfrak{m}\cdot M=\langle f_1,\ldots,f_q\rangle\cdot M$, where $\langle f_1,\ldots,f_q\rangle$ denotes the ideal generated by the components $f_i \in {}_n\mathcal{O}$ of F.

Preparation Theorem: *The ${}_n\mathcal{O}$-module M is finitely generated over ${}_q\mathcal{O}$ if and only if M is finitely generated over ${}_n\mathcal{O}$ and $M/{}_q\mathfrak{m}M$ is a finite dimensional complex vector space. In this case the elements $u_1,\ldots,u_r \in M$ generate M over ${}_q\mathcal{O}$ if and only if their residue classes generate the complex vector space $M/{}_q\mathfrak{m}M$.*

One direction (if M is finitely generated over ${}_q\mathcal{O}$) and the last statement follow from the corollary above. The other direction follows from the Division Theorem in §1 : We use the notation introduced there and prove first the Preparation Theorem for the projection $\mathbb{C}^n \to \mathbb{C}^{n-1}$, namely:

Lemma: *Let M be a finitely generated ${}_n\mathcal{O}$-module so that $\dim_{\mathbb{C}} M/{}_{n-1}\mathfrak{m}M<\infty$. Then M is finitely generated as ${}_{n-1}\mathcal{O}$-module.*

Proof: There are generators $u_1,\ldots,u_r$ of M as ${}_n\mathcal{O}$-module whose residue classes generate $M/{}_{n-1}\mathfrak{m}M$. Every $x \in M$ can be written as

(1) $$x = c_1u_1+\ldots+c_ru_r$$

with $c_i \in {}_n\mathcal{O}$ and $c_i(0,t)=\gamma_i \in \mathbb{C}$. We apply this to $x=tu_j$ and obtain

(2) $$\sum_{i=1}^{r}(c_{ji}-\delta_{ji}t)u_i = 0 \text{ with } c_{ji}\in {}_n\mathcal{O} \text{ and } c_{ji}(0,t)=\gamma_{ji}\in\mathbb{C}.$$

Let D denote the $r\times r$ matrix with elements $d_{ji}=c_{ji}-\delta_{ji}t \in {}_n\mathcal{O}$ and let Γ denote the $r\times r$ matrix with elements γ_{ji}. Then $D(0,t)=\Gamma-tE$, where E denotes the unit matrix. Let

$d = \det D \in {}_n\mathcal{O}$. We obtain: $d(0,t)$ is a polynomial in t of degree r with highest coefficient ± 1 . In particular, d is regular in t of order $s \leq r$. By general matrix theory there is a matrix B with elements in ${}_n\mathcal{O}$ so that $B \cdot D = d \cdot E$. Therefore by (2)

$$(3) \qquad d \cdot u_j = 0 \quad \text{for } j=1,\ldots,r .$$

We return to an arbitrary x as in (1) and divide each coefficient c_i by d using the Division Theorem. Then

$$(4) \qquad x = p_1 u_1 + \ldots + p_r u_r \quad ,$$

where each p_i is a polynomial in t of degree $<s\leq r$ with coefficients in ${}_{n-1}\mathcal{O}$. Here (3) has of course been used. The presentation (4) says: As an ${}_{n-1}\mathcal{O}$-module M is generated by the elements $t^j u_i$ for $j=0,\ldots,s-1$ and $i=1,\ldots,r$.

Proof of the Preparation Theorem: We decompose F into

$$F: \mathbb{C}^n \xrightarrow{G} \mathbb{C}^{q+n} \xrightarrow{\pi_1} \mathbb{C}^{q+n-1} \xrightarrow{\pi_2} \ldots \xrightarrow{\pi_n} \mathbb{C}^q$$

where $G(z)=(F(z),z)$ and where the π_i are the projections which omit the last coordinate. It suffices to prove the theorem for G and for the π_i . For G it is trivial because $G^o: {}_{q+n}\mathcal{O} \to {}_n\mathcal{O}$ is epimorphic, for each π_i the theorem is the lemma which we have just proved.

Actually the Division Theorem is equivalent to the Preparation Theorem: Let $f \in {}_n\mathcal{O}$ be regular of order r with respect to t . We apply the Preparation Theorem to $M={}_n\mathcal{O}/\langle f\rangle$ and the projection $\mathbb{C}^n \to \mathbb{C}^{n-1}$ onto the first coordinates. The complex vector space $M/{}_{n-1}\mathfrak{m}M = {}_n\mathcal{O}/\langle z_1,\ldots,z_{n-1},f\rangle = {}_n\mathcal{O}/\langle z_1,\ldots,z_{n-1},t^r\rangle$ has a base consisting of the residue classes of $1,t,\ldots,t^{r-1} \in {}_n\mathcal{O}$. Therefore M is generated over ${}_{n-1}\mathcal{O}$ by the residue classes of these elements in M: Every $h \in {}_n\mathcal{O}$ can be written as $h = q \cdot f + a_{r-1}t^{r-1} + \ldots + a_o$ with $a_i \in {}_{n-1}\mathcal{O}$ and $q \in {}_n\mathcal{O}$.

The Preparation Theorem is also true for real C^{∞}-maps (Malgrange). It is one of the basic tools for the investigation of singularities of C^{∞}-mappings, see Golubitsky / Guillemin: Stable Mappings and their Singularities. Springer-Verlag New York 1973.

§6 Finite Maps

Algebraic Preliminaries: Let R and S denote commutative rings with unit elements. Let $R \subset S$ be a ring extension. An element $s \in S$ is called *integral* over R if there is a polynomial $p(z) \in R[z]$ with highest coefficient 1 such that $p(s)=0$. Polynomials with highest coefficient 1 will be called *monic*. The extension $R \subset S$ is called integral, if every $s \in S$ is integral over R. The extension $R \subset S$ is called *finite*, if S is finitely generated as R-module.

Proposition 1: *Let R be noetherian. Every finite ring extension $R \subset S$ is integral.*

Proof: Let $s \in S$ be given. We consider the R-submodule $R[s] \subset S$. Since R is noetherian and S is finitely generated, $R[s]$ is also a finitely generated R-module. Let n be the highest degree occuring in a set of generators. Then $s^{n+1}=b_n s^n+\ldots+b_o$ for suitable $b_\nu \in R$.

A holomorphic germ $\pi: (X,0) \to (\mathbb{C}^k,0)$ is called *finite* (at 0) if the ring extension $\pi^o({}_k\mathcal{O}_o) \subset \mathcal{O}_{X,o}$ is finite. If $(Y,0) \subset (X,0)$, then $\pi|Y$ is finite if π is.

We consider a slightly more general situation. Let $F: (\mathbb{C}^n,0) \to (\mathbb{C}^k,0)$ be a holomorphic mapping germ and let $\mathfrak{a} \subset {}_n\mathcal{O}$ be an ideal. We use the following notations:

$$\Phi: {}_k\mathcal{O} \xrightarrow{F^o} {}_n\mathcal{O} \twoheadrightarrow {}_n\mathcal{O}/\mathfrak{a} = S \quad , \qquad R = \Phi({}_k\mathcal{O}) \quad .$$
$$f \longmapsto \tilde{f}$$

The mapping germ F is called *finite with respect to* $\mathfrak{a}$ if $R \subset S$ is a finite ring extension. In this case F is also finite with respect to every $\mathfrak{b} \supset \mathfrak{a}$. If $(X,0) \subset (\mathbb{C}^n,0)$ is the germ of an analytic set, then $\pi = F|X$ is finite at 0 if and only if F is finite with respect to the ideal $\mathfrak{J}(X) \subset {}_n\mathcal{O}_o$.

Minimal Polynomials: Results on finite maps rely on the investigation of minimal polynomials. We consider a mapping germ $F: (\mathbb{C}^n,0) \to (\mathbb{C}^k,0)$ which is finite with respect to the ideal $\mathfrak{a} \subset {}_n\mathcal{O}$. We use the notations of the previous paragraph. Since R is the homomorphic image of the noetherian ring ${}_k\mathcal{O}$ it is also noetherian. We apply proposition 1: For every $\tilde{f} \in S$ there is a monic poynomial $\mu(z) \in R[z]$ such that $\mu(\tilde{f}) = 0$. If μ has minimal possible degree it is called a minimal polynomial of $\tilde{f}$. There is a monic polynomial $M(z) \in {}_k\mathcal{O}[z]$ which is transformed into $\mu(z)$ by means of $\Phi: {}_k\mathcal{O} \twoheadrightarrow R$. If μ is minimal, M is called a minimal polynomial of $\tilde{f}$, too. Let us assume that $f \in {}_n\mathfrak{m}$. We define $H: (\mathbb{C}^n,0) \to (\mathbb{C}^k \times \mathbb{C},0)$ by $H(\xi) = (F(\xi), f(\xi))$, and consider $\chi: {}_{k+1}\mathcal{O} \xrightarrow{H^o} {}_n\mathcal{O} \twoheadrightarrow S$. A monic polynomial $M \in {}_k\mathcal{O}[z] \subset {}_{k+1}\mathcal{O}$ is a minimal polynomial of $\tilde{f}$ if and only if $\chi(M) = 0$ and M has minimal degree among all monic polynomials P with $\chi(P) = 0$.

Proposition 2: *If* $M \in {}_k\mathcal{O}[z]$ *is a minimal polynomial of some* $\tilde{f} \in S$ *with* $f \in {}_n\mathfrak{m}$, *then* M *is a Weierstrass polynomial.*

Proof: Since M is monic it is regular with respect to z of some order $s \le r = \text{degree } M$. We have $s = r$ if and only if M is a Weierstrass polynomial. If $s < r$, then by the Vorbereitungssatz $M = u \cdot M^*$ for a unit $u \in {}_{k+1}\mathcal{O}$ and a Weierstrass polynomial $M^* \in {}_k\mathcal{O}[z]$ of degree $s < r$. Then $0 = \chi(M) = \chi(u) \cdot \chi(M^*)$. Since $\chi(u)$ is a unit, this implies $\chi(M^*) = 0$, a contradiction to the minimal degree r .

Let V be a connected neighbourhood of 0 in $\mathbb{C}^k$. Then

$$\mathcal{O}(V) \to {}_k\mathcal{O}_0 \quad , \quad g \mapsto \text{germ of } g \text{ at } 0$$

is a monomorphism. We consider $\mathcal{O}(V)$ as a subalgebra of ${}_k\mathcal{O} = {}_k\mathcal{O}_0$. If a germ ${}_k\mathcal{O}$ belongs to $\mathcal{O}(V)$ we say that f *is defined on* V . Given finitely many $g_1,\ldots,g_m \in {}_k\mathcal{O}$, there is a V , such that all g_i are defined on V .

For a polynomial $P \in \mathcal{O}(V)[z_1,\ldots,z_m]$ we denote by $P(\eta;z_1,\ldots,z_m) \in \mathbb{C}[z_1,\ldots,z_m]$ the polynomial which is obtained from P by evaluating the coefficients at $\eta \in V$. Obviously P determines the holomorphic function $V \times \mathbb{C}^n \to \mathbb{C}$, $(\eta;t_1,\ldots,t_m) \mapsto P(\eta,t) = P(\eta;t_1,\ldots,t_m)$.
Let $z_i \in {}_n\mathfrak{m}$ denote the i-th coordinate function. Let $M_i \in {}_k\mathcal{O}[z_i]$ be a minimal polynomial for $\tilde{z}_i \in S$. We obtain the following

<u>Corollary 3</u> (to proposition 2): *For every neighbourhood* U *of* 0 *in* $\mathbb{C}^n$ *there is a neighbourhood* V *of* 0 *in* $\mathbb{C}^k$ *such that the coefficients of* M_i *for* $i=1,\ldots,n$ *are defined on* V *and such that*

$$\{(\eta,\xi) \in V \times \mathbb{C}^n : M_1(\eta,\xi_1) = \ldots = M_n(\eta,\xi_n) = 0\} \subset V \times U .$$

Proof: Since M_i is a Weierstrass polynomial, $M(0,\xi_i) = \xi_i^{r_i}$ for some r_i . The corollary follows because the roots of a polynomial depend continuously on the coefficients.

<u>Theorem 4</u> (Topological properties of finite maps): *Let* X *be an analytic set with* $0 \in X$. *Let* $F: X \to \mathbb{C}^k$ *be a holomorphic map with* $F(0)=0$ *which is finite at* 0 . *There is a neighbourhood* W *of* 0 *in* X *such that the restriction* $\pi: F|W$ *satisfies:*

(a) *Every fibre* $\pi^{-1}(\eta)$ *is a finite set.*

(b) *For every smaller neighbourhood* $W_o \subset W$ *of* 0 *in* X *there is a neighbourhood* V_o *of* 0 *in* $\mathbb{C}^k$ *such that* $\pi^{-1}(V_o) \subset W_o$.

Observe, that (a) and (b) remain true if W is replaced by a smaller W' .

Proof: We use the minimal polynomials $M_i \in \mathcal{O}(V)[z_i]$ of the coordinate functions z_i of $\mathbb{C}^n$. Their coeffincients are defined in a neighbourhood V of 0 in $\mathbb{C}^k$. We choose W connected and so small that $F(W) \subset V$. Then $\xi \mapsto M_i(F(\xi), z_i(\xi))$ is a well defined holomorphic function on W . Its germ at 0 vanishes, therefore $M_i(F(\xi), z_i(\xi)) = 0$ for all $\xi \in W$. Let r_i=degree of M_i . For every $\eta \in V$ there are at most r_i different roots of $M_i(\eta, z_i) = 0$. Therefore there are at most $r_i \cdots r_n$ different points $\xi \in W$ with $F(\xi) = \eta$. This proves (a) . For (b) let U_o be a neighbourhood of 0 in $\mathbb{C}^n$ so that $W_o = X \cap U_o$. By the preceding corollary we obtain

$\{(\eta,\xi) \in V_o \times \mathbb{C}^n : M_1(\eta,\xi_1) = \ldots = M_n(\eta,\xi_n) = 0\} \subset V_o \times U_o$ if V_o is small enough. This implies $\pi^{-1}(V_o) \subset W_o$ because $M_i(\pi(\xi), z_i(\xi)) = 0$.

The following two lemmas will be employed in §7. Let $y_1, \ldots, y_k, z_1, \ldots, z_n$ denote the coordinate functions in $\mathbb{C}^k \times \mathbb{C}^n = \mathbb{C}^{n+k}$. We consider ${}_k\mathcal{O}$ and ${}_n\mathcal{O}$ as subalgebras of ${}_{n+k}\mathcal{O}$ consisting of the germs of those functions which depend on the first k respectively the last n coordinates only. In particular there is the polynomial subalgebra ${}_k\mathcal{O}[z_1, \ldots, z_n] \subset {}_{k+n}\mathcal{O}$.

<u>Lemma 5</u>: *Let* $M_i \in {}_k\mathcal{O}[z_i]$ *be monic polynomials,* $i = 1, \ldots, n$. *For every* $f \in {}_{k+n}\mathcal{O}$ *there is a* $p \in {}_k\mathcal{O}[z_1, \ldots, z_n]$ *such that* $f - p$ *is contained in the ideal generated by* $M_1, \ldots, M_n$.

Proof by induction on n beginning with the trivial case $n=0$: Using the division theorem f is divided by M_n with remainder f' . Then $f-f'$ is a multiple of M_n and $f' \in {}_{k+n-1}\mathcal{O}[z_n]$. By induction hypothesis for every coefficient a_ν of f' there is a polynomial $p_\nu \in {}_k\mathcal{O}[z_1,\ldots,z_{n-1}]$ such that $a_\nu - p_\nu$ lies in the ideal generated by $M_1,\ldots,M_{n-1}$. We replace in f' each a_ν by p_ν and obtain a polynomial $p \in {}_k\mathcal{O}[z_1,\ldots,z_n]$ such that $f'-p$ lies in the ideal generated by $M_1,\ldots,M_{n-1}$. Since $f-f'$ is a multiple of M_n the lemma follows.

Lemma 6: *Let* $F: (\mathbb{C}^n,0) \to (\mathbb{C}^k,0)$ *be finite with respect to* $\mathfrak{a} \subset {}_n\mathcal{O}$. *Then* $S=R[\tilde{z}_1,\ldots,\tilde{z}_n]$. (We use the notations introduced in the beginning of this section. The $z_1,\ldots,z_n$ denote the coordinate functions of $\mathbb{C}^n$) .

Proof: Let $M_i \in {}_k\mathcal{O}[z_i]$ denote a minimal polynomial of $\tilde{z}_i$. We apply lemma 5: For every $f \in {}_{k+n}\mathcal{O}$ there is a $p \in {}_k\mathcal{O}[z_1,\ldots,z_n]$ such that $f-p$ lies in the ideal generated by $M_1,\ldots,M_n$. We define $s: (\mathbb{C}^n,0) \to (\mathbb{C}^{k+n},0)$ by $s(\xi)=(F(\xi),\xi)$ and consider the induced epimorphism $\sigma: {}_{k+n}\mathcal{O} \xrightarrow{s^0} {}_n\mathcal{O} \twoheadrightarrow S$. Then $\sigma(g)=\Phi(g)$ for $g \in {}_k\mathcal{O}$ and $\sigma(f)=\tilde{f}$ for $f \in {}_n\mathcal{O}$. Since $\sigma(M_i)=0$ we obtain $\sigma(f)=\sigma(p) \in R[\tilde{z}_1,\ldots\tilde{z}_n]$ for every $f \in {}_{k+n}\mathcal{O}$. This implies $S=R[\tilde{z}_1,\ldots,\tilde{z}_n]$, because σ is surjective.

The identity $id_{\mathbb{C}^n}$ is finite with respect to every ideal $\mathfrak{a} \subset {}_n\mathcal{O}$. But this trivial example would not justify the theory developed so far. We are rather interested in finite maps $F: (\mathbb{C}^n,0) \to (\mathbb{C}^k,0)$ with $k<n$. They can be obtained as follows:

Proposition 7: *Given an ideal* $\langle 0\rangle \neq \mathfrak{a} \subset {}_n\mathcal{O}$ *the coordinates can be changed linearly so that the projection* $\pi: \mathbb{C}^n \to \mathbb{C}^{n-1}$ *onto the first* $n-1$ *coordinates is finite with respect to* $\mathfrak{a}$.

Proof: We choose a $0 \neq p \in \mathfrak{a}$ and transform the coordinates linearly such that p becomes regular of some order r with respect to the last coordinate. We denote the coordinates by

$z_1,\dots,z_{n-1},t$. We use the notation $\Phi: {}_{n-1}\mathcal{O} \xrightarrow{\pi^o} {}_n\mathcal{O} \twoheadrightarrow {}_n\mathcal{O}/\mathfrak{m} = S$ and $R=\Phi({}_{n-1}\mathcal{O})$ which has been introduced earlier in this section and claim that S is generated as R-module by $1,\tilde{t},\dots,\tilde{t}^{r-1}$. Let $\tilde{f}\in S$ be given. We divide by p using the division theorem, $f=q\cdot p+a_{r-1}t^{r-1}+\dots+a_o$ with $a_\nu \in {}_{n-1}\mathcal{O}$. Since $p\in\mathfrak{a}$ this implies $\tilde{f}=\tilde{a}_{r-1}\tilde{t}^{r-1}+\dots+\tilde{a}_o$ with $\tilde{a}_\nu\in R$.

For the repeated application of proposition 7 we need the

Composition of Finite Maps: *Let* $F: (\mathbb{C}^n,0) \to (\mathbb{C}^k,0)$ *be finite with respect to* $\mathfrak{a}\subset {}_n\mathcal{O}$, *and let* $G: (\mathbb{C}^k,0) \to (\mathbb{C}^q,0)$ *be finite with respect to* $\mathfrak{b}$ = *kernel of* ${}_k\mathcal{O} \xrightarrow{F^o} {}_n\mathcal{O} \twoheadrightarrow {}_n\mathcal{O}/\mathfrak{a}$. *Then* $G\circ F: (\mathbb{C}^n,0) \to (\mathbb{C}^q,0)$ *is finite with respect to* $\mathfrak{a}$, *furthermore* ${}_q\mathcal{O} \xrightarrow{G^o} {}_k\mathcal{O} \twoheadrightarrow {}_k\mathcal{O}/\mathfrak{b}$ *and* ${}_q\mathcal{O} \xrightarrow{G^o} {}_k\mathcal{O} \xrightarrow{F^o} {}_n\mathcal{O} \twoheadrightarrow {}_n\mathcal{O}/\mathfrak{a}$ *have the same kernel.*

The mapping germ $F: (\mathbb{C}^n,0) \to (\mathbb{C}^k,0)$ is called *strict* with respect to the ideal $\mathfrak{a}\subset {}_n\mathcal{O}$, if $\Phi: {}_k\mathcal{O} \xrightarrow{F^o} {}_n\mathcal{O} \twoheadrightarrow {}_n\mathcal{O}/\mathfrak{a}$ is monomorphic.

Theorem 8 (Existence of finite and strict maps): *Given an ideal* $\mathfrak{a}\subset {}_n\mathcal{O}$ *the coordinates can be changed linearly so that for some* $k\le n$ *the projection* $\pi: \mathbb{C}^n \to \mathbb{C}^k$ *onto the first* k *coordinates is finite and strict with respect to* $\mathfrak{a}$.

The proof is by induction on n . We assume that the theorem holds true for $n-1$ variables. If $\mathfrak{a}=\langle 0\rangle\subset {}_n\mathcal{O}$, there is nothing to prove because then $id_{\mathbb{C}^n}$ is finite and strict. If $\mathfrak{a}\neq 0$, after a linear change of coordinates the projection $\pi_1: \mathbb{C}^n \to \mathbb{C}^{n-1}$ onto the first $n-1$ coordinates becomes finite with respect to $\mathfrak{a}$. Let $\mathfrak{b}$= kernel of $\Phi: {}_{n-1}\mathcal{O} \to {}_n\mathcal{O} \twoheadrightarrow {}_n\mathcal{O}/\mathfrak{a}$. By induction hypothesis there is a linear change of coordinates in $\mathbb{C}^{n-1}$ such that for some $k\le n-1$ the projection $\pi_2: \mathbb{C}^{n-1} \to \mathbb{C}^k$ onto the first k coordinates is finite and strict with respect to $\mathfrak{b}$. Then the composed projection $\pi=\pi_2\circ\pi_1$ is finite and strict with respect to $\mathfrak{a}$.

In §9, proposition 8 we shall see that $k = \dim N(\mathfrak{a})$. There is much more known about the topology of finite and strict maps than the results of theorem 4. This will be the topic of the following section. But more algebra will be involved and the proofs become more complicated.

§7 Finite and Strict Maps

The Discriminant: Let R be an entire ring (commutative, with unit element), let $C = \operatorname{Quot} R$ denote its quotient field. Let $P \in R[z]$ be a monic polynomial. There is a finite extension of fields $C \subset K$ such that P splits into linear factors $P(z)=(z-z_1)\dots(z-z_r)$ with $z_\nu \in K$ and r = degree of P . A priori the so called *discriminant*

$$d = \prod_{i \neq j} (z_i - z_j)$$

is an element of K . But d is symmetric with respect to the z_i . Therefore d can be written as a polynomial in the elementary symmetric functions of the z_i with coefficients in $\mathbb{Z}$. The elementary symmetric functions are the coefficients of P , they belong to R and so $d \in R$. By definition P has a multiple root if and only if $d=0$.

A prototype of a finite and strict map: Let $U \subset \mathbb{C}^k$ be an open and connected neighbourhood of 0 . Let $P \in \mathcal{O}(U)[z]$ be a Weierstrass polynomial of degree r . We consider the analytic set

$$\Sigma = \Sigma_P = \{(\eta,t) \in U \times \mathbb{C} : P(\eta,t)=0\}$$

together with the projection $\pi: \Sigma \to U$, $\pi(\eta,t)=\eta$. We assume that the discriminant $d \in \mathcal{O}(U)$ of P is $\neq 0$. Otherwise the following result would be empty.

Proposition 1: *Every point* $\eta \in U$ *with* $d(\eta) \neq 0$ *has a neighbourhood* V *in* U *such that* $\pi^{-1}(V)$ *is the disjoint union of* r *open sets* $V_1, \ldots, V_r$ *in* Σ, *each of which is mapped biholomorphically onto* V *by means of* π.

Proof: Since $d(\eta) \neq 0$ there are r simple roots $t_1, \ldots, t_r$ of $P(\eta, z) = 0$. The partial derivative $\partial P / \partial z$ is $\neq 0$ at each (η, t_i). By the Implicit Function Theorem there are holomorphic functions $\tau_i \colon V \to \mathbb{C}$ defined in a neighbourhood V of η such that $\tau_i(\eta) = t_i$ and $P(y, \tau_i(y)) = 0$ for all $y \in V$. Then $P(y,z) = (z - \tau_1(y)) \ldots (z - \tau_r(y))$ for $y \in V$. If V is sufficiently small, $d(y) \neq 0$ for $y \in V$ and so $\tau_i(y) \neq \tau_j(y)$ for $i \neq j$. Let $V_i = \{(y,t) \in V \times \mathbb{C} \colon t = \tau_i(y)\} \subset \Sigma$. The V_i are disjoint, and π maps V_i biholomorphically onto V because $y \mapsto (y, \tau_i(y))$ is the inverse map.

Remark: The projection π is strict and finite at 0. 'Strict' follows from the continuity of roots: For every neighbourhood U of 0 in $\mathbb{C}$ there is a neighbourhood V of 0 in $\mathbb{C}^k$ such that for every $\eta \in V$ there is a $t \in U$ with $P(\eta, t) = 0$. 'Finite' follows by dividing an arbitrary $h \in {}_{k+1}\mathcal{O}_0$ by the germ of P using the Division Theorem, §6 proposition 7.

Let $F \colon (\mathbb{C}^n, 0) \to (\mathbb{C}^k, 0)$ be a mapping germ which is strict and finite with respect to a *prime* ideal $\mathfrak{g} \subset {}_n\mathcal{O}$. Let X be an analytic set in $\mathbb{C}^n$ whose germ at 0 is $N(\mathfrak{g})$. The restriction $F|X$ looks essentially like the projection $\pi \colon \Sigma_P \to U \subset \mathbb{C}^k$ which has been described above. In order to make this statement precise and to define the corresponding polynomial P we need further

Algebraic Preliminaries: We consider a finite extension $R \subset S$ of entire rings. The quotient fields are denoted by $C = \text{Quot } R$ and $E = \text{Quot } S$. There is the induced extension of fields. We assume furthermore that R is noetherian.

Proposition 2: (a) *Let* S *be generated by* $x_1,\ldots,x_n \in S$ *as* R-*algebra, i.e.* $S=R[x_1,\ldots,x_n]$. *Then* $E=C[x_1,\ldots,x_n]$.
(b) *Let* S *be generated by* $s_1,\ldots,s_m \in S$ *as* R-*module. Then* E *is generated by* $s_1,\ldots,s_m$ *as vector space over* C . *The elements* $s_1,\ldots,s_m$ *form a base of* E *over* C *if and only if they generate* S *freely as* R-*module.*

Proof: Every $x \in C[x_1,\ldots,x_n] \subset E$ can be written as a polynomial in $x_1,\ldots,x_n$ with coefficients in C . There is a common denominator $r \in R, \neq 0$, of the coefficients. Then $rx \in R[x_1,\ldots,x_n] \subset S$. By proposition 1 in §6 $(rx)^q + b_{q-1}(rx)^{q-1}+\ldots+b_o=0$ for some q and certain $b_\nu \in R$, thus x is algebraic over C . Therefore $C(x)=C[x] \subset C[x_1,\ldots,x_n]$, particularly $x^{-1} \in C[x_1,\ldots,x_n]$ if $x \neq 0$. Thus $C[x_1,\ldots,x_n]$ is a field which contains S , hence $C[x_1,\ldots,x_n]=E$. This proves (a). Obviously $S=R[s_1,\ldots,s_m]$, hence $E=C[s_1,\ldots,s_m]$ by (a). Thus every $x \in E$ is a linear combination of monomials $s_1^{r_1}\ldots s_m^{r_m}$ with coefficients in C. Every monomial is a linear combination of $s_1,\ldots,s_m$ with coefficients in R , thus x is a linear combination of $s_1,\ldots,s_m$ with coefficients in C . Assume that $\alpha_1 s_1+\ldots+\alpha_m s_m=0$ for certain $\alpha_i \in C$. Let $0 \neq r \in R$ be a common denominator of the α_i . Then $r\alpha_1 \cdot s_1+\ldots+r\alpha_m \cdot s_m=0$ and $r\alpha_i \in R$. If S is freely generated by $s_1,\ldots,s_m$ this implies $r\alpha_i=0$ hence $\alpha_i=0$ for every i . The converse is quite trivial.

If $C \subset E$ is a finite extension of fields, every $x \in E$ has a unique minimal polynomial $M \in C[z]$ which is characterized by

(1) $M(x) = 0$,

(2) M has minimal degree among all $P \in C[z]$ with $P(x)=0$,

(3) M is monic.

We can replace (2) by (2') : M is irreducible.

Proposition 3: *Let* Quot $R = C \subset E$ *be a finite extension of fields. Let* $x \in E$ *be integral over* R . *Then the coefficients of its minimal polynomial* $M \in C[z]$ *are also integral over* R. See Lang, p.240, for the proof.

Corollary 4: *Let* $R \subset S$ *satisfy the general hypotheses. Let* R *be integrally closed in* C *. The minimal polynomial of every* $x \in S \subset E$ *belongs to* $R[z] \subset C[z]$.

"R integrally closed in C" means that every $y \in C$ which is integral over R belongs to R . This is e.g. the case if R is factorial, see Lang, p.240.

Let $C \subset E$ be a field extension, char C = 0. An element $x \in E$ is called *primitive* for $C \subset E$ if $E = C(x)$.

Proposition 5 (on the primitive element): Let $x_1,..,x_n \in E$ be such that $E = C(x_1,..,x_n)$ is a finite extension of C . For almost all n-tupels $\alpha = (\alpha_1,..,\alpha_n) \in C^n$ the element $x = \alpha_1 x_1 + ... + \alpha_n x_n$ is primitive for $C \subset E$. For a proof see Lang, p.185 f.

We consider the situation of corollary 4. Proposition 5 implies that the extension $C \subset E$ has a primitive element $x \in S \subset E$. Let $M \in R[z]$ be its minimal polynomial, and let $d \in R$ be the discriminant of M . The degree of M is the degree $[E:C]$ of the extension.

Proposition 6: *In the situation which has just been described, for every* $s \in S$ *there is a polynomial* $T \in R[z]$ *of degree strictly less than* $[E:C]$ *such that* $d \cdot s = T(x)$. For a proof see v.d.Waerden II, p.80/81.

Since $s = T(x)/d$, the discriminant d is also called a *universal denominator*. Every minimal polynomial is irreducible and thus has no multiple roots so that $d \neq 0$.

We return to a holomorphic germ $F: (\mathbb{C}^n,0) \to (\mathbb{C}^k,0)$ which is finite and strict with respect to a prime ideal $\mathfrak{g} \subset {}_n\mathcal{O}$. We use the notation which has been introduced in the beginning of §6. We identify ${}_k\mathcal{O}$ with R by Φ. Since $\mathfrak{g}$ is prime, S is entire. Since $R \cong {}_k\mathcal{O}$ is factorial, it is integrally closed in $C = \mathrm{Quot}\, R$. Let $E = \mathrm{Quot}\, S$. We use the preceding algebraic preliminaries. The extension $C \subset E$ is finite. Let

$$r = [E:C]$$

denote its degree. Since $S = R[\tilde{z}_1, \ldots, \tilde{z}_n]$ by lemma 6 in §6, we have $E = C[\tilde{z}_1, \ldots, \tilde{z}_n]$ by proposition 2(a). So a suitable linear combination ζ of the $\tilde{z}_i$ with coefficients in C is a primitive element for $C \subset E$ (Proposition 5). After a linear change of the coordinates we may assume that $\zeta = \tilde{z}_1$. The minimal polynomial $M_1 \in {}_k\mathcal{O}[z_1]$ of $\tilde{z}_1$ is monic, irreducible (in $C[z_1]$) and has degree r. Its discriminant $d \in {}_k\mathcal{O}$ is $\neq 0$, because M_1 is irreducible. There is a connected neighbourhood V of 0 in $\mathbb{C}^k$ such that the coefficients of M_1, hence also d, are defined on V, thus $M_1 \in \mathcal{O}(V)[z_1]$ and $d \in \mathcal{O}(V)$. We define the hypersurface

$$\Sigma = \{(\eta,t) \in V \times \mathbb{C} \colon M_1(\eta,t) = 0\}$$

Let $\mathfrak{g} = \langle p_1, \ldots, p_m \rangle$. There is a connected neighbourhood U of 0 in $\mathbb{C}^n$ such that $p_1, \ldots, p_m$ are defined on U. If U is small enough, F is also defined on U and $F(U) \subset V$. We define furthermore

$$H\colon U \to V \times \mathbb{C}, \quad H(\xi) = (F(\xi), z_1(\xi))$$
$$X = N(p_1, \ldots, p_m) \subset U, \quad X' = \{\xi \in X \colon d \circ F(\xi) \neq 0\}$$
$$\Sigma' = \{(\eta,t) \in \Sigma \colon d(\eta) \neq 0\}, \quad \pi = F|X .$$

<u>Main Theorem 7</u>: *Let* $F\colon (\mathbb{C}^n,0) \to (\mathbb{C}^k,0)$ *be finite and strict with respect to the prime ideal* $\mathfrak{g} \subset {}_n\mathcal{O}$. *There are neighbourhoods* U *of* 0 *in* $\mathbb{C}^n$ *and* V *of* 0 *in* $\mathbb{C}^k$ *such that the following holds true:*

(a) $H(X) \subset \Sigma$; (b) H *maps* X' *biholomorphically onto* Σ' .

Proof: As in the proof of lemma 6 in §6 we consider $s\colon (\mathbb{C}^n,0) \to (\mathbb{C}^{k+n},0)$, $s(\xi) = (F(\xi),\xi)$ and its induced epimorphism $\sigma\colon {}_{k+n}\mathcal{O} \xrightarrow{s^o} {}_n\mathcal{O} \twoheadrightarrow {}_n\mathcal{O}/\mathfrak{g} = S$. By proposition 6 (on the universal denominator) there are polynomials $T_i \in {}_k\mathcal{O}[z_1]$ of degree $<r$ such that $\sigma(d\cdot z_i - T_i)=0$. We choose especially. $T_1(z_1)=d\cdot z_1$.

Lemma 8: *Let* $Q \subset {}_k\mathcal{O}$ *be a subalgebra which contains the coefficients of* M_1 *and of* $T_1,\ldots,T_n$. *Let* $\mathfrak{n}' \subset Q[z_1,\ldots,z_n]$ *denote the ideal which is generated by* M_1, $d\cdot z_2-T_2,\ldots,$ $d\cdot z_n-T_n$. *Let* $f \in Q[z_1,\ldots,z_n]$ *satisfy* $\sigma(f)=0$. *Then* $d^m\cdot f \in \mathfrak{n}'$ *for* $m=$degree f .

Remark: $Q={}_k\mathcal{O}$ and $=\mathcal{O}(V)$ for a connected neighbourhood V of 0 are the interesting cases.

Proof: There is a $f^* \in Q[z_1,\ldots,z_n]$ such that the substitution $z_i \mapsto d\cdot z_i$ transforms f^* into $d^m\cdot f$. The substitution $z_i \mapsto T_i$ transforms f^* into some $q \in Q[z_1]$, and $d^m\cdot f-q$ belongs to the ideal generated by $d\cdot z_2-T_2,\ldots,$ $d\cdot z_n-T_n$. Since M_1 is monic, we divide q in $Q[z_1]$ by M_1 such that the remainder $q' \in Q[z_1]$ has degree $<r$. Since $\sigma(M_1)=0$ we have $\sigma(q)=\sigma(q')$. On the other hand $\sigma(q)=\sigma(d^m f)$ $=d^m\sigma(f)=0$ because $\sigma(d\cdot z_i-T_i)=0$. Therefore q' is a polynomial of degree $<r$ which has $\tilde{z}_1$ as a zero. But r is the minimal degree of such a polynomial $\neq 0$ so that $q'=0$. Hence q is a multiple of M_1 and $d^m\cdot f$ belongs to $\mathfrak{n}'$.

Lemma 9: *Let* $\mathfrak{n} \subset {}_{k+n}\mathcal{O}$ *denote the ideal which is generated by* M_1, $d\cdot z_2-T_2$, $\ldots$, $d\cdot z_n-T_n$. *Let* $f \in {}_{k+n}\mathcal{O}$ *satisfy* $\sigma(f)=0$. *Then* $d^q\cdot f \in \mathfrak{n}$ *for some* q .

Proof: Let $M_i \in {}_k\mathcal{O}[z_i]$ be the minimal polynomial of $\tilde{z}_i$. Then $\sigma(M_i)=0$ so that $d^m M_i \in \mathfrak{n}' \subset {}_k\mathcal{O}[z_1,\ldots,z_n]$ for some m because of lemma 8. According to lemma 5 in §6 there is a $f' \in {}_k\mathcal{O}[z_1,\ldots,z_n]$ such that $f-f'$ lies in the ideal generated by $M_1,\ldots,M_n$, therefore $d^m(f-f') \in \mathfrak{n}$ and $\sigma(f')=\sigma(f)=0$. Hence by lemma 8 $d^r f' \in \mathfrak{n}' \subset \mathfrak{n}$ for some r so that finally $d^q f \in \mathfrak{n}$ for $q=\max\{m,r\}$.

The First Choice of V: The neighbourhood V of 0 in $\mathbb{C}^k$ is chosen to be connected and so small that the coefficients of M_i and T_i for $i=1,\dots,n$ are defined on V. By lemma 8 applied to $Q=\mathcal{O}(V)$ there is an exponent α such that

(1) $d^{\alpha}\cdot M_i \in$ ideal in $\mathcal{O}(V)[z_1,\dots,z_n]$ generated by M_1 and $d\cdot z_i-T_i$ for $i=2,\dots,n$.

The First Choice of U: The neighbourhood U of 0 in $\mathbb{C}^n$ is chosen to be connected, so that the generators $p_1,\dots p_m$ of $\mathfrak{g}$ are defined on U, so that F is defined on U, and so that $F(U)\subset V$. Then $H_i\colon U \to V\times\mathbb{C}$, $H_i(\xi)=(F(\xi),z_i(\xi))$ is well defined. (H_1 is the H of the theorem). The germs of every $M_i\circ H_i$ and of every $(d\circ F)\cdot z_i-T_i\circ H$ at 0 belong to $\mathfrak{g}$. We shrink U so that

(2) $M_i\circ H_i$ and $(d\circ F)\cdot z_i-T_i\circ H$ belong to the ideal $\mathfrak{P}=\langle p_1,\dots,p_m\rangle$ in $\mathcal{O}(U)$.

The Second Choice of U and V: Let U and V be of first choice. Then $\varphi_i\colon V\times U\to\mathbb{C}$, $\varphi_i(\eta,\xi)=f_i(\xi)-y_i(\eta)$, is defined. Here y_i denotes the i-th coordinate function of $V\subset\mathbb{C}^k$ and $f_i=y_i\circ F$, $i=1,\dots,k$. We consider $\mathcal{O}(U)\subset\mathcal{O}(V\times U)$ as subalgebra, particularly $p_j\in\mathcal{O}(V\times U)$. We apply lemma 9 to the germs of φ_i and p_j at 0 and obtain an exponent γ such that $d^{\gamma}\cdot\varphi_i$ and $d^{\gamma}\cdot p_j$ belong to the ideal $\mathfrak{m}\subset {}_{k+n}\mathcal{O}$. We can now shrink U and V such that they remain of first choice and satisfy in addition:

(3) There is an exponent γ such that for every $f\in\{p_1,\dots,p_m,\varphi_1,\dots,\varphi_k\}$ the product $d^{\gamma}\cdot f$ belongs to the ideal in $\mathcal{O}(U\times V)$ which is generated by M_1, and $d\cdot z_i-T_i$ for $i=2,\dots,n$.

The Final Choice of U and V: Let U and V be of second choice. By corollary 3 of §6 can be shrinked such that

(4) $\{(\eta,\xi)\in V\times\mathbb{C}^n\colon M_1(\eta,\xi_1)=\dots=M_n(\eta,\xi_n)=0\}\subset V\times U$.

We replace U by $U=F^{-1}(V)$.

The End of the Proof: We claim that the Main Theorem holds true for the neighbourhoods U and V. We consider $M_1 \circ H$ and obtain from (2) that $H(X) \subset \Sigma$. Let

$$V' = \{\eta \in V : d(\eta) \neq 0\} .$$

We define G by its components g_i as follows

$$G: V' \times \mathbb{C} \to \mathbb{C}^n,\ g_i(\eta,t) = \frac{1}{d(\eta)} T_i(\eta,t) \quad \text{for} \quad i=1,\dots,n ,$$

particularly $g_1(\eta,t)=t$. Then

(5) $$G \circ H|X' = id_{X'}$$

because of (2). We have $M_i(\eta,g_i(\eta,t))=0$ for $(\eta,t) \in \Sigma'$ because of (1), therefore $G(\Sigma') \subset U$. Using (3) for the φ_i we obtain $F \circ G(\eta,t)=\eta$ for $(\eta,t) \in \Sigma'$, in particular $G(\Sigma') \subset U$. We use (3) again, this time for p_j and see that $p_j \circ G(\eta,t)=0$ for $(\eta,t) \in \Sigma'$, i.e. $G(\Sigma') \subset X$. The definition of H together with $F \circ G(\eta,t)=\eta$ implies

(6) $$H \circ G|\Sigma' = id_{\Sigma'} .$$

According to (5) and (6) the maps $H|X': X' \to \Sigma'$ and $G|\Sigma': \Sigma' \to X'$ are inverse to one another.

Remarks: (i) The statements (a) and (b) of the Main Theorem remain true if V is replaced by a smaller neighbourhood V_o of 0 provided U is replaced by $U_0=F^{-1}(V_0)$.

(ii) If we want to shrink U to a smaller neighbourhood U' of 0 so that (a) and (b) remain true, we first choose a neighbourhood V_o of 0 in V with $\pi^{-1}(V_0) \subset U'$ using Theorem 4(b) of §6. We replace U by $U_0 = F^{-1}(V_0) \cap U'$. Then (a) and (b) hold true for U_0, V_0 instead of U,V.

(iii) There is a compact ball B with centre 0 in U. By Theorem 4(b) of §6 there is a neighbourhood V_o of 0 in V with $\pi^{-1}(V_0) \subset B$. By (i) the statements (a) and (b) remain true if U and V are replaced by $U_0=F^{-1}(V_0)$ and V_0. By this restriction we obtain furthermore:

(c) $$\pi|(X \cap U_0) \to V_0 \quad \text{is surjective and proper} .$$

'proper' means: For every compact $K \subset V_0$ the inverse image $\pi^{-1}(K)$ is compact. Indeed, $\pi^{-1}(K)$ is closed in X and contained in B, hence closed in $X \cap B$. But $X \cap B$ is compact. In order to prove surjectivity we consider Σ' first. For every $\eta \in V_0$ with $d(\eta) \neq 0$ there is a t such that $(\eta, t) \in \Sigma'$ and so by (b), there is a $\xi \in X$ with $\pi(\xi) = \eta$. If $\eta \in V_0$ and $d(\eta) = 0$ there is a sequence $\{\eta_\nu\}$ in V_0 with $d(\eta_\nu) = 0$ which converges to η. For every η_ν there is a $\xi_\nu \in X$ with $\pi(\xi_\nu) = \eta_\nu$. Since $\pi | (X \cap U_0)$ is proper a subsequence of $\{\xi_\nu\}$ converges to some $\xi \in X \cap U_0$, and by continuity $\pi(\xi) = \eta$.

Summary: For later reference we summarize some of the results of this and the preceding section. Let X be an analytic set with $0 \in X$ and let $\pi: X \to \mathbb{C}^k$ be a holomorphic map with $\pi(0) = 0$. We describe the structure of π in the neighbourhood of 0 under the following conditions:

X is irreducible at 0.

π is strict and finite at 0, i.e. $\pi^o: {}_k\mathcal{O}_o \hookrightarrow \mathcal{O}_{X,o}$ is injective and a finite ring extension.

For simplicity we shall write ${}_k\mathcal{O} = {}_k\mathcal{O}_o$ and $\mathcal{O}_X = \mathcal{O}_{X,o}$. Let $f_1, \ldots, f_k$ denote the components of π. The complex vector space $\mathcal{O}_X / \langle f_1, \ldots, f_k \rangle = \mathcal{O}_X / \pi^o({}_k\mathfrak{m}) \cdot \mathcal{O}_X$ has finite dimension s. This s is the minimal number of generators of $\mathcal{O}_X$ as ${}_k\mathcal{O}$-module, see the Corollary to Nakayama's lemma in §5.

Since both ${}_k\mathcal{O}$ and $\mathcal{O}_X$ are entire, there are the quotient fields ${}_k\mathcal{M} = \mathrm{Quot}\ {}_k\mathcal{O}$ and $\mathcal{M}_X = \mathrm{Quot}\ \mathcal{O}_X$. They are called the fields of germs of meromorphic functions. The ring extension π^o induces a field extension ${}_k\mathcal{M} \hookrightarrow \mathcal{M}_X$ of finite degree r. By Proposition 2

$$r = [\mathcal{M}_X : {}_k\mathcal{M}] \leq \dim \mathcal{O}_X / \langle f_1, \ldots, f_k \rangle = s$$

with equality if and only if $\mathcal{O}_X$ is a free ${}_k\mathcal{O}$-module.

There are neighbourhoods W of 0 in X and V of 0 in $\mathbb{C}^k$ such that the restriction $\pi: W \to V$ is surjective, proper, and has finite fibres. For every neighbourhood W_o of 0 in

W there is a neighbourhood V_0 of 0 in V with $\pi^{-1}(V_0) \cap W \subset W_o$. This follows from Theorem 4 in §6 and Remark (iii) above. Finally we combine (b) of the Main Theorem with Proposition 1: There is a $0 \neq d \in \mathcal{O}(V)$ such that every point $\eta \in V$ with $d(\eta) \neq 0$ has an open neighbourhood Ω in V for which $\pi^{-1}(\Omega) \cap W = \Omega_1 \cup \ldots \cup \Omega_r$ is the disjoint union of r open sets Ω_i each of which is mapped biholomorphically onto Ω by π, briefly: Outside the set $\{\eta \in V: d(\eta)=0\}$ the restriction $\pi: W \to V$ is a *smooth holomorphic covering with* r *sheets.*

§8 The Nullstellensatz

The Nullstellensatz $\mathfrak{J}N(\mathfrak{a}) = \mathrm{rad}\,\mathfrak{a}$ follows from the Main Theorem 7 of the preceding section. The special case $N(\mathfrak{a}) = \{0\}$ enables us to characterize finite (and strict) maps topologically.

<u>Theorem 1</u> (Rückert's Nullstellensatz): *For every ideal* $\mathfrak{a} \subset {}_n\mathcal{O}$ *we have* $\mathfrak{J}N(\mathfrak{a}) = \mathrm{rad}\,\mathfrak{a}$.

Proof: We begin with a prime ideal $\mathfrak{a} = \mathfrak{g}$. Here it suffices to show that $\mathfrak{J}N(\mathfrak{g}) \subset \mathfrak{g}$. By theorem 6 in §6 there is an $F: (\mathbb{C}^n,0) \to (\mathbb{C}^k,0)$ which is finite and strict with respect to $\mathfrak{g}$. We consider ${}_k\mathcal{O}$ as subalgebra of $S = {}_n\mathcal{O}/\mathfrak{g}$. Let $f \in \mathfrak{J}N(\mathfrak{g})$. By proposition 6 of §7 (universal denominator) there is a polynomial $T \in {}_k\mathcal{O}[z_1]$ of degree $< r$ such that $d \cdot \tilde{f} = T(\tilde{z}_1)$; here $r = [\mathrm{Quot}\, S : \mathrm{Quot}\, {}_k\mathcal{O}]$. There are neighbourhoods U of 0 in $\mathbb{C}^n$ and V of 0 in $\mathbb{C}^k$ such that the Main Theorem 7 of §7 holds true, such that f is defined on U, the coefficients of T are defined on V, and such that f and $(d \circ F) \cdot f - T \circ H$ vanish on X.

$\mathfrak{p} \subset \mathcal{J}N(\mathfrak{p})$. We conclude that $T \circ H|X = 0$. For every $\eta \in V' = \{y \in V: d(y) \neq 0\}$ there are r different complex numbers $t_1,\ldots,t_r$ such that $M_1(\eta,t_i)=0$, i.e. $(\eta,t_i) \in \Sigma'$. By the Main Theorem (b) there are $\xi_1,\ldots,\xi_r \in X$ such that $H(\xi_i)=(\eta,t_i)$. Then $0=T \circ H(\xi_i)=T(\eta,t_i)$. But $T(\eta,z_1)$ has a degree $< r$, therefore $T(\eta,z_1)=0$ for every $\eta \in V'$. Since V' is dense in V , we conclude $T(\eta,z_1)=0$ for every $\eta \in V$, i.e. T is the zero-polynomial in ${}_k\mathcal{O}[z_1]$. Then $d \cdot \tilde{f}=0$ in S , and so $\tilde{f}=0$ in S because S is entire and $d \neq 0$.

Since ${}_n\mathcal{O}$ is noetherian, an arbitrary ideal $\mathfrak{a}$ has a representation $\mathfrak{a} = \mathfrak{q}_1 \cap \ldots \cap \mathfrak{q}_s$ so that $\mathfrak{p}_i = \operatorname{rad} \mathfrak{q}_i$ is prime (Primary decomposition of Lasker-Noether, see e.g. Lang p.152) Then $N(\mathfrak{a}) = N(\mathfrak{q}_1) \cup \ldots \cup N(\mathfrak{q}_s) = N(\mathfrak{p}_1) \cup \ldots \cup N(\mathfrak{p}_s)$ and $\mathcal{J}N(\mathfrak{a}) = \mathcal{J}N(\mathfrak{p}_1) \cap \ldots \cap \mathcal{J}N(\mathfrak{p}_s)$. By the Nullstellensatz for prime ideals, $\mathcal{J}N(\mathfrak{a}) = \mathfrak{p}_1 \cap \ldots \cap \mathfrak{p}_s = \operatorname{rad} \mathfrak{q}_1 \cap \ldots \cap \operatorname{rad} \mathfrak{q}_s = \operatorname{rad} \mathfrak{a}$.

Proposition 2: *Let* $\mathfrak{a} \subset {}_n\mathfrak{m}$. *The* $\mathbb{C}$-*algebra* ${}_n\mathcal{O}/\mathfrak{a}$ *is finite dimensional if and only if* $N(\mathfrak{a}) = \{0\}$.

Proof: We show that both conditions are equivalent to

(*) There is a q such that $\mathfrak{m}^q \subset \mathfrak{a}$.

Assume (*). Then $\mathcal{O}/\mathfrak{m}^q \to \mathcal{O}/\mathfrak{a}$ is epimorphic. This implies $\dim \mathcal{O}/\mathfrak{a} < \infty$ because $\dim \mathcal{O}/\mathfrak{m}^q < \infty$. Furthermore $\{0\}=N(\mathfrak{m}^q) \supset N(\mathfrak{a})$.

Assume $\dim \mathcal{O}/\mathfrak{a} < \infty$. There is the sequence of sub-vector spaces $\mathcal{O}/\mathfrak{a} \supset (\mathfrak{m}+\mathfrak{a})/\mathfrak{a} \supset (\mathfrak{m}^2+\mathfrak{a})/\mathfrak{a} \supset \ldots$. The dimensions cannot drop infinitely many times, therefore $(\mathfrak{m}^q+\mathfrak{a})/\mathfrak{a} = (\mathfrak{m}^{q+1}+\mathfrak{a})/\mathfrak{a}$ for some q, in other words: $(\mathfrak{m}^q+\mathfrak{a})/\mathfrak{a} = \mathfrak{m} \cdot (\mathfrak{m}^q+\mathfrak{a})/\mathfrak{a}$. By Nakayama's lemma, $(\mathfrak{m}^q+\mathfrak{a})/\mathfrak{a} = 0$, i.e. $\mathfrak{m}^q \subset \mathfrak{a}$.

Assume $N(\mathfrak{a})=\{0\}$. By the Nullstellensatz, $\operatorname{rad}\mathfrak{a} = \mathcal{J}(\{0\}) = \mathfrak{m}$. For every coordinate function z_i there is a q_i such that $z_i^{q_i} \in \mathfrak{a}$ and so $\mathfrak{m}^q \subset \mathfrak{a}$ for $q \geq q_1+\ldots+q_n$.

Theorem 3 (Topological Characterization of Finite Maps): *Let* $X \subset U$ *be an analytic set where* $U \subset \mathbb{C}^n$ *is open. For every*

$x \in X$ *let* $\mathfrak{m}_x \subset {}_n\mathcal{O}_x$ *denote an ideal such that* $N(\mathfrak{m}_x)$ *is the germ of* X *at* x. *Let* $F: U \to \mathbb{C}^k$ *be a holomorphic map, and let* $\pi = F|X$.

The germ of F *at* $c \in X$ *is finite for* $\mathfrak{m}_c$ *if and only if* c *is an isolated point of the fibre* $\pi^{-1}(\pi(c))$. *In this case for all* x *in a neighbourhood* W *of* c *in* X *the germ of* F *at* x *is finite for* $\mathfrak{m}_x$.

Proof: One direction (if F is finite with respect to $\mathfrak{m}_c$) is Theorem 4(a) in §6. Vice versa let c be isolated in $\pi^{-1}(\pi(c))$. The $N(\mathfrak{m}_c + \langle f_1,\dots,f_k\rangle) = \{c\}$. Here $\langle f_1,\dots,f_k\rangle$ denotes the ideal in ${}_n\mathcal{O}_c$ which is generated by the compo - nents $f_1,\dots,f_k$ of F. By Proposition 2 ${}_n\mathcal{O}_c/(\mathfrak{m}_c + \langle f_1,\dots,f_k\rangle)$ is finite dimensional. Then ${}_n\mathcal{O}_c/\mathfrak{m}_c$ is finitely generated as ${}_k\mathcal{O}_{\pi(c)}$-module because of the Preparation Theorem, see §5. The last statement of the theorem follows because every x in a neighbourhood W of c in X is also isolated in its fibre according to Theorem 4(a) in §6.

Proposition 4 (Strict Maps): *We consider the same situation as in the first paragraph of the preceding Theorem 3. The following statements are equivalent for a point* $c \in X$:

(a) F *is strict with respect to* $\mathfrak{m}_c$.

(b) F *is strict with respect to* $\mathcal{J}(X_c)$, *where* X_c *denotes the germ of* X *at* c.

(c) F *is strict with respect to* $\mathcal{J}(X^*)$ *for at least one irreducible component* X^* *of* X_c.

Proof: (a) implies (b): We have $\mathcal{J}(X_c) = \operatorname{rad}\mathfrak{m}_c$ according to the Nullstellensatz. It suffices to show that the compo- sition ${}_k\mathcal{O} \xrightarrow{\Phi} {}_n\mathcal{O}/\mathfrak{m} \to {}_n\mathcal{O}/\operatorname{rad}\mathfrak{m}$ is injective. (We have omitted the indices c and $\pi(c)$). Assume that f is mapped to 0. Then $f \circ F \in \operatorname{rad}\mathfrak{m}$, i.e. $f^q \circ F \in \mathfrak{m}$ for some q. Since Φ is injective, $f^q = 0$ and so $f = 0$.

(b) implies (c): Let $X_c = X_1 \cup \dots \cup X_s$ be the decomposition into irreducible components. If (c) were not true, there would be a $0 \neq f_i \in {}_k\mathcal{O}$ with $f_i \circ \pi | X_i = 0$ for every i. Then $(f_1 \ldots f_s) \circ \pi = 0$, thus by (b) $f_1 \ldots f_s = 0$, i.e. $f_i = 0$ for some i, a contradiction.

(c) implies (a): If (a) is not true, there is a $0 \neq f \in {}_k\mathcal{O}$ with $f \circ F \in \mathfrak{n}$ and so $f \circ \pi = 0$. But then $f \circ \pi | X_i = 0$ for every irreducible component.

<u>Theorem 5</u> (Topological Characterization of Finite and Strict Maps): *We consider the situation which has been described in the first paragraph of theorem 3. The germ* F *at* $c \in X$ *is finite and strict with respect to the ideal* $\mathfrak{n}_c$ *if and only if* c *is isolated in its fibre* $\pi^{-1}(\pi(c))$, *and* π *is open at* c, *i.e. for every neighbourhood* W *of* c *in* X *the image* $\pi(W)$ *is a neighbourhood of* $\pi(c)$ *in* $\mathbb{C}^k$.

Proof: 'Finite' and 'isolated' are equivalent to one another by theorem 3. 'Strict and 'finite' implies 'open at c' : By proposition 4 there is a component X^* of X_c such that F is strict with respect to $\mathcal{J}(X^*)$, and also finite because $\mathfrak{n}_c \subset \mathcal{J}(X^*)$. It suffices to show that $\pi | X^*$ is open at c. Thus we may assume that X is irreducible at c. We apply the Main Theorem 7 of §7 to the prime ideal $\mathcal{J}(X_c)$. By the remark (c) following the proof of this theorem the image $\pi(X)$ contains a neighbourhood V of 0 in $\mathbb{C}^k$. By theorem 4(b) of §6 there is a smaller neighbourhood V_1 of 0 in V such that $\pi^{-1}(V_1) \subset W$. Then $V_1 \subset \pi(W)$.

'open at c' implies 'strict': Assume that for some $f \in {}_k\mathcal{O}_c$ we have $f \circ F \in \mathfrak{n}_c$. The $f \circ \pi = 0$ in a neighbourhood W of c in X and so $f | \pi(W) = 0$. Since $\pi(W)$ is a neighbourhood of 0 in $\mathbb{C}^k$ this implies $f = 0$.

§9 The Dimension

For a vector space V the dimension dim V can be characterized as the minimal number n of linear functions $f_i: V \to \mathbb{C}$ such that $N(f_1,\dots,f_n)=\{x\in V: f_1(x)=\dots=f_n(x)\}=\{0\}$. Similarly the dimension $\dim_o X$ of an analytic set X at the point 0 is defined to be the minimal number n of germs $f_1,\dots,f_n\in \mathcal{O}_{X,o}$ such that $N(f_1,\dots,f_n) = \{x\in X: f_1(x)=\dots=f_n(x)\} = \{0\}$. Of course $\dim_o X$ depends only on the germ of X at 0 .

Proposition 1 (Algebraic description of the dimension): *Let* $\mathfrak{a}\subset {}_n\mathcal{O}$ *be an ideal, let* $S = {}_n\mathcal{O}/\mathfrak{a}$. *The dimension* $k = \dim N(\mathfrak{a})$ *is the minimal number of elements* $\varphi_1,\dots,\varphi_k\in S$ *such that* $S/\langle\varphi_1,\dots,\varphi_k\rangle$ *is a finite dimensional complex vector space.*

Proof: By definition $\dim N(\mathfrak{a})\le k$ if and only if there are $f_1,\dots,f_k\in {}_n\mathcal{O}$ so that $N(\mathfrak{a}+\langle f_1,\dots,f_k\rangle) = \{0\}$. By proposition 2 in §8 $N(\mathfrak{a}+\langle f_1,\dots,f_k\rangle) = \{0\}$ if and only if ${}_n\mathcal{O}/(\mathfrak{a}+\langle f_1,\dots,f_k\rangle)$ is finite dimensional. The proposition follows because ${}_n\mathcal{O}\to S$ induces an isomorphism ${}_n\mathcal{O}/(\mathfrak{a}+\langle f_1,\dots,f_k\rangle)\cong S/\langle\varphi_1,\dots,\varphi_k\rangle$ for the images φ_i of f_i .

The definition of the dimension has some immediate consequences like

(1) If $(Y,0)\subset(X,0)$ then $\dim_o Y\le \dim_o X$

(2) $\dim_o N(f_1,\dots,f_q)\ge \dim_o X-q$ for any $f_1,\dots,f_q\in\mathcal{O}_{X,o}$

(3) $\dim_x \mathbb{C}^n\le n$ for every $x\in\mathbb{C}^n$.

These results will be improved below, e.g. $\dim_x\mathbb{C}^n=n$. The topological characterization of finite maps (theorem 3 in §8) implies

Proposition 2: *We have* $\dim_o X\le k$ *if and only if there is a finite germ* $F: (X,0)\to(\mathbb{C}^k,0)$. *In this case* $\dim_x X\le k$ *for every* x *in a neighbourhood of* 0 .

Further results on the dimension follow from the

Active Lemma 3: *Let* $\mathfrak{a} \subset {}_n\mathcal{O}$ *be an ideal. If* $f \in {}_n\mathcal{O}$ *is not a zero divisor modulo* $\mathfrak{a}$ *then* $\dim N(\mathfrak{a}+\langle f\rangle) = \dim N(\mathfrak{a})-1$.

Proof: Let $Y = N(\mathfrak{a}+\langle f\rangle)$ and let $k = \dim N(\mathfrak{a})$. It suffices to prove $\dim Y \leq k-1$ because of (1) and (2). There is an $F: (\mathbb{C}^n,0) \to (\mathbb{C}^k,0)$ which is finite with respect to $\mathfrak{a}$. Let $M \in {}_k\mathcal{O}[z]$ be a minimal polynomial of $\tilde{f} \in {}_n\mathcal{O}/\mathfrak{a}$,

$$M(z) = z^p + a_{p-1}z^{p-1} + \ldots + a_o \quad , \quad a_i \in {}_k\mathcal{O} .$$

Then

$$f\cdot(f^{p-1}+(a_{p-1}\circ F)f^{p-2}+\ldots+a_1\circ F) = -(a_o\circ F) \quad \text{modulo } \mathfrak{a} .$$

Since f is not a zero divisor modulo $\mathfrak{a}$, this implies $a_o \neq 0$. There are coordinates $y_1,\ldots,y_k$ of $\mathbb{C}^k$ such that a_o is regular with respect to y_k , i.e. $a_o(0,\ldots,0,y_k) = y_k^r\cdot u(y_k)$ with $u(0)\neq 0$ for some r . Let $f_1,\ldots,f_k$ denote the components of F . We claim that $Z = N(f_1|Y,\ldots,f_{k-1}|Y) = \{0\}$. Indeed, $0 \in Z$ and if $\xi \in Z$ then $a_o(F(\xi)=0$ because $\xi \in Y$; further $f_k(\xi)^k\cdot u(f_k(\xi)) = a_o(F(\xi))$ because $f_1(\xi)=\ldots=f_{k-1}(\xi)=0$. This implies $f_k(\xi)=0$ and so $F(\xi)=0$. Since F is finite with respect to $\mathfrak{a}$ and $\xi \in N(\mathfrak{a})$ we obtain $\xi=0$ by theorem 4(a) in §6.

Corollary 4: *If* $(X,0)$ *is irreducible, if* $(Y,0) \subset (X,0)$, ***and*** *if* $\dim_o Y = \dim_o X$, *then* $(Y,0) = (X,0)$.

Proof: Assume that $(Y,0) \neq (X,0)$. There is an $0 \neq f \in \mathcal{O}_{X,o}$ such that $Y \subset N(f)$. Since X is irreducible, f is not a zero divisor, therefore $\dim Y \leq \dim N(f) = \dim X-1$, the last equality follows from the Active Lemma.

The Active Lemma yields by induction on n :

Theorem 5: $\dim_x \mathbb{C}^n = n$ *for every* $x \in \mathbb{C}^n$.

We define $\dim\inf_o X = \min\{\dim_o X_i : X_i$ irreducible component of X at 0 } .

Proposition 6: *For* $f_1,\dots,f_q \in {}_n\mathcal{O}$ *we have* $\dim \inf N(f_1,\dots, f_q) \geq n-q$.

Proof: Let X^* be an irreducible component of $X=N(f_1,\dots,f_q)$ at 0 with $\dim X^* = \dim \inf_0 X$. Let U be a neighbourhood of 0 in $\mathbb{C}^n$ such that $f_1,\dots,f_q$ are defined on U and $\dim_x X^* \leq \dim_0 X^* < \dim_0 X$ for every $x \in X^* \cap U$. There is an $x \in X^* \cap U$ which does not belong to any other component of X , therefore $\dim_x X^* = \dim_x X$. Let k be this dimension. There are $g_1,\dots,g_k \in {}_n\mathcal{O}_x$ such that $N(g_1|X,\dots,g_k|X) = \{x\}$. If $f_1,\dots,f_q$ denote also the germs at x , we obtain $N(f_1,\dots, f_q,g_1,\dots,g_k) = \{x\}$ therefore $q+k \geq n = \dim_x \mathbb{C}^n$. This implies $\dim \inf_0 X = \dim_0 X^* \geq k \geq n-q$.

Proposition 7 (Characterization of Hypersurfaces): *The germ* $(X,0) \subset (\mathbb{C}^n,0)$ *of an analytic set is a hypersurface if and only if* $\dim \inf X = n-1$.

Proof: One direction (if $X=N(f)$ is a hypersurface) is a particular case of the preceding proposition. Assume now that $\dim \inf X = n-1$ and let X be irreducible. The prime ideal $\mathcal{J}(X)$ contains an irreducible f . We claim that $\langle f \rangle = \mathcal{J}(X)$. Otherwise there would be a $g \in \mathcal{J}(X)$ which is not a zero divisor modulo $\langle f \rangle$, and so by the Active Lemma $\dim X \leq \dim N(f,g) = \dim N(f)-1 = n-2$, a contradiction. In general the assumption $\dim \inf X = n-1$ implies $\dim \inf X_i = n-1$ for every irreducible component, therefore $\mathcal{J}(X_i) = \langle f_i \rangle$ for some irreducible f_i . Let $X = X_1 \cup \dots \cup X_r$. Then $N(f_1 \cdots f_r) = N(f_1) \cup \dots \cup N(f_r) = X_1 \cup \dots \cup X_r = X$.

Proposition 2 characterizes $k = \dim_0 X$ as the smallest number for which a finite germ $F\colon (X,0) \to (\mathbb{C}^k,0)$ exists. This F must be strict because of proposition 7 and the composition of finite maps in §6. Vice versa, if F is finite and strict, then $\dim_0 X=k$. We prove a bit more:

Proposition 8: *If* $F\colon (X,0) \to (\mathbb{C}^k,0)$ *is finite and strict, there is an irreducible component* X^* *of* X *at* 0 *with* $\dim X^* = k$.

Proof: By Proposition 4 in §8 there is an irreducible

component X^* such that $F|X^*$ is strict and finite. By the end of the summary of §7 there are points $x \in X^*$ arbitrary close to 0 such that a neighbourhood of x in X is isomorphic to an open set in $\mathbb{C}^k$, therefore $\dim_x X^* = k$. Since $\dim_x X^* \leq \dim_o X \leq k$ by proposition 2, we obtain $\dim_o X^* = k$.

Corollary 9:

$$\dim_o X = \max \{\dim_o X_i : X_i \text{ irreducible component of } X \text{ at } 0\}$$

Let $\gamma(\mathfrak{n})$ denote the minimal number of generators of the ideal $\mathfrak{n} \subset {}_n\mathfrak{m}$, let $\delta(\mathfrak{n}) = n-\gamma(\mathfrak{n})$. There are the following inequalities:

(4) $\delta(\mathfrak{n}) \leq \dim \inf N(\mathfrak{n}) \leq \dim N(\mathfrak{n}) \leq \operatorname{edim} \mathfrak{n}$.

The first one is proposition 6, the middle one follows from (1), the last one follows from §4: We may assume that $\mathfrak{n}$ is minimally embedded, i.e. $\operatorname{edim} \mathfrak{n} = n$. Then $N(\mathfrak{n}) \subset \mathbb{C}^n$ implies $\dim N(\mathfrak{n}) \leq n$ by (1) and (3).

Proposition 10: *The following statements are equivalent:*

(5) $\mathfrak{n}$ *is regular*.

(6) $\delta(\mathfrak{n}) = \operatorname{edim} \mathfrak{n}$.

(7) $\dim N(\mathfrak{n}) = \operatorname{edim} \mathfrak{n}$.

Proof: For "(5) implies (6)" see the end of §4. "(6) implies (7)" follows from (4). "(7) implies (5)": We may assume that $\mathfrak{n}$ is minimally embedded. Then $N(\mathfrak{n}) = \mathbb{C}^n$ by corollary 4, hence $\mathfrak{n} = \langle 0 \rangle$.

§10 Annihilators

This section and the following one present some methods of commutative algebra which will be used in §12 for the study of complete intersections.

Let M be a module over the commutative ring R with unit. An element $\alpha \in R$ is called M-*annihilator* if $\alpha x=0$ for some $0 \neq x \in M$. For every fixed $x \in M$ the set

$$\text{Ann } x = \{\alpha \in R: \alpha x=0\}$$

is an ideal. It is called the annihilator ideal of x.

Lemma 1: *A maximal annihilator ideal is prime.*

Proof: Let $\text{Ann } x$ be maximal, let $\alpha\beta \in \text{Ann } x$ but $\beta \notin \text{Ann } x$. Then $\beta x \neq 0$ and $\alpha \in \text{Ann}(\beta x)$. Obviously $\text{Ann } x \subset \text{Ann}(\beta x)$ and by maximality $\text{Ann } x = \text{Ann}(\beta x)$.

A prime ideal $\mathfrak{p}$ is called *associated* to M if $\mathfrak{p} = \text{Ann } x$ for some $0 \neq x \in M$.

Observation: *In this case* $\text{Ann } x = \text{Ann } \alpha x$ *for every* $\alpha \in R$ *with* $\alpha x \neq 0$.

Let $\text{Ass } M$ denote the set of all prime ideals which are associated to M.

Lemma 2: *If* $0 \to U \to M \xrightarrow{\pi} Q$ *is exact, then* $\text{Ass } U \subset \text{Ass } M \subset \text{Ass} U \cup \text{Ass } Q$.

Proof: The first inclusion is obvious. Consider some prime $\text{Ann } x$ for $0 \neq x \in M$. If $U \cap Rx \neq 0$ there is an α such that $0 \neq \alpha x \in U$ and $\text{Ann } x = \text{Ann } \alpha x \in \text{Ass } U$ by the observation. If $U \cap Rx = 0$, the submodule $Rx \subset M$ is mapped isomorphically by π onto the submodule $R\pi(x) \subset Q$, therefore $\text{Ann } x = \text{Ann } \pi(x) \in \text{Ass } Q$.

Corollary 3: $\text{Ass}(U \oplus Q) = \text{Ass } U \cup \text{Ass } Q$.

Lemma 4: *Let* $x_1,\dots,x_n \in M$ *be such that* $\mathrm{Ann}\, x_1, \dots, \mathrm{Ann}\, x_n$ *are pairwise different prime ideals. The submodule generated by* $x_1,\dots,x_n$ *is a direct sum* $Rx_1+\dots+Rx_n = Rx_1 \oplus \dots \oplus Rx_n$ *and* $\mathrm{Ass}(Rx_1+\dots+Rx_n) = \{\mathrm{Ann}\, x_1, \dots, \mathrm{Ann}\, x_n\}$.

Proof by induction on n: The observation implies $\mathrm{Ass}\, Rx_1 = \{\mathrm{Ann}\, x_1\}$. The induction hypothesis $\mathrm{Ass}(Rx_1+\dots+Rx_{n-1}) = \{\mathrm{Ann}\, x_1, \dots, \mathrm{Ann}\, x_{n-1}\}$ and the observation imply $(Rx_1+\dots+Rx_{n-1}) \cap Rx_n = 0$ because $\mathrm{Ann}\, x_n \neq \mathrm{Ann}\, x_i$ for $i \neq n$. Therefore $Rx_1+\dots+Rx_{n-1}+Rx_n = (Rx_1+\dots+Rx_{n-1}) \oplus Rx_n$ and $\mathrm{Ass}(Rx_1+\dots+Rx_n) = \mathrm{Ass}(Rx_1+\dots+Rx_{n-1}) \cup \{\mathrm{Ann}\, x_n\}$ by corollary 3.

Theorem 5: *Let* R *be noetherian and* M *finitely generated. The set of all* M*-annihilators is the union of the associated prime ideals. There are only finitely many associated prime ideals.*

Proof: Obviously the elements of the associated prime ideals are annihilators. Vice versa: Every annihilator lies in some Ann x with $x \neq 0$. Since R is noetherian, we may assume that Ann x is maximal. Then $\mathrm{Ann}\, x \in \mathrm{Ass}\, M$ by lemma 1. If Ass M were infinite, there would be an infinitely increasing sequence of submodules $Rx_1 \subsetneq Rx_1+Rx_2 \subsetneq \dots$ because of lemma 4. This is impossible because of the hypotheses.

Proposition 6: *Let* R *be noetherian and* M *finitely generated. If all elements of an ideal* $\mathfrak{a} \subset R$ *are* M*-annihilators, then* $\mathfrak{a}x=0$ *for some* $0 \neq x \in M$.

Proof: By theorem 5 $\mathrm{Ass}\, M = \{\mathfrak{p}_1,\dots,\mathfrak{p}_n\}$ is finite and $\mathfrak{a} \subset \mathfrak{p}_1 \cup \dots \cup \mathfrak{p}_n$. This implies $\mathfrak{a} \subset \mathfrak{p}_j$ for some j , see Proposition 8 below. Now $\mathfrak{p}_j = \mathrm{Ann}\, x$ for some $0 \neq x \in M$, hence $\mathfrak{a}x=0$.

Proposition 7: *Let* R *be noetherian and let* $M_1,\dots,M_n$ *be finitely generated* R*-modules. Let* $\mathfrak{a} \subset R$ *be an ideal. If for each* $i=1,\dots,n$ *not all elements of* $\mathfrak{a}$ *are* M_i*-annihilators, there is a common* $\alpha \in \mathfrak{a}$ *which is not an* M_i*-annihilator for every* $i=1,\dots,n$.

Proof: By proposition 6 **there is an** $\alpha \in \mathfrak{a}$ **which is not an** annihilator for $M_1 \oplus \ldots \oplus M_n$ and therefore not an annihilator for each M_i .

Proposition 8: *Let the ideal* $\mathfrak{a}$ *be contained in the union* $\mathfrak{p}_1 \cup \ldots \cup \mathfrak{p}_n$ *of finitely many prime ideals. Then* $\mathfrak{a} \subset \mathfrak{p}_j$ *for some* $j=1,\ldots,n$.

Proof: We may assume that for $i \neq j$ we have $\mathfrak{p}_i \not\subset \mathfrak{p}_j$. Otherwise we could delete $\mathfrak{p}_i$. Assume now that $\mathfrak{a}$ is contained in no $\mathfrak{p}_j$. Then for any j the ideal $\mathfrak{a}'_j = \mathfrak{a} \cap \bigcap_{i \neq j} \mathfrak{p}_i$ is not contained in $\mathfrak{p}_j$: If $a \in \mathfrak{a} \setminus \mathfrak{p}_j$ and $p_i \in \mathfrak{p}_i \setminus \mathfrak{p}_j$, the product $a \cdot \prod_{i \neq j} p_i$ is in $\mathfrak{a}'_j$ but not in $\mathfrak{p}_j$ because $\mathfrak{p}_j$ is prime. Let now $a_j \in \mathfrak{a}'_j \setminus \mathfrak{p}_j$. Then $\sum_j a_j \in \mathfrak{a}$ without being in any $\mathfrak{p}_i$. This contradicts $\mathfrak{a} \subset \bigcup_j \mathfrak{p}_j$.

§11 Regular Sequences

Let R be a noetherian local ring with maximal ideal $\mathfrak{m}$. Let M be an R-module. An element $\alpha \in R$ is called M-*active* if $\alpha \in \mathfrak{m}$ and α is not an M-annihilator.

Lemma 1: Let M be finitely generated, let $\mathfrak{a} \subset R$ be an ideal and let $\alpha, \beta \in \mathfrak{a}$ be M-active. There are $M/\alpha M$-active elements in $\mathfrak{a}$ if and only if there are $M/\beta M$-active elements in $\mathfrak{a}$.

Proof: Assume that all elements of $\mathfrak{a}$ are $M/\alpha M$-annihilators. There is an $x \in M, \notin \alpha M$ such that $\mathfrak{a} x \subset \alpha M$ because of proposition 6 in the preceding section, particularly $\beta x = \alpha y$ for some $y \in M$. Here $y \notin \beta M$ because $y = \beta z$ would imply $\beta x = \alpha\beta z$, hence $x = \alpha z \in \alpha M$ because β is active. On the other hand $\alpha\mathfrak{a} y = \beta\mathfrak{a} x \subset \beta\alpha M$. Thus for every $\xi \in \mathfrak{a}$ there is a $u \in M$ such that $\alpha\xi y = \beta\alpha u$. Since α is active this implies $\xi y = \beta u$, i.e.

ξ is an $M/\beta M$-annihilator.

A sequence $(\alpha_1,\dots,\alpha_n)$ in R is called *M-regular,* if α_{i+1} is active for $M_i=M/\langle\alpha_1,\dots,\alpha_i\rangle M$, $i=0,\dots,n-1$ and $M=M_0$. Obviously $(\beta_1,\dots\beta_m)$ is M_n-regular if and only if $(\alpha_1,\dots,\alpha_n,\beta_1,\dots,\beta_m)$ is M-regular.

Lemma 2 (Commuting Lemma): *Let* M *be finitely generated. If* (α,β) *is M-regular then so is* (β,α) .

Proof: (1) β is M-active: Assume the contrary. Then there is a $y\neq 0$ such that $\beta y=0$. Since β is $M/\alpha M$-active, $\beta y=0$ implies $y=\alpha y'$ for some y' . Then $\beta y'=0$ because $0=\beta y=\alpha\beta y'$ and α is M-active. We repeat the argumentation with y' instead of y and obtain a y'' such that $y'=\alpha y''$ and $\beta y''=0$, etc. Thus we obtain an infinite sequence $y,y',y'',\dots,y^{(n)},\dots$ with $y=\alpha^n y^{(n)}$. Then $0\neq y\in N=\bigcap_{n=0}^{\infty}\alpha M^n$; but $\alpha N=N$, hence $N=0$ by Nakayama's lemma.

(2) α is $M/\beta M$-active: Otherwise $\alpha x=\beta y$ for some $x\in M\setminus\beta M$ and some $y\in M$. Since β is $M/\alpha M$-active, $y=\alpha z$ for some $z\in M$, thus $\alpha x=\alpha\beta z$, and so $x=\beta z$ because α is M-active. This contradicts $x\notin\beta M$.

Lemma 3: *Let* M *be finitely generated. If there is a regular M-sequence of length* n *in the ideal* $\mathfrak{a}\subset R$, *there is such a sequence which begins with a prescribed* α, *if* α *is M-active.*

Proof by induction on n : The beginning $n=1$ is trivial. The case $n=2$ is lemma 1. In order to go from $n-1$ to n we assume that $(\alpha_1,\dots,\alpha_n)$ in $\mathfrak{a}$ is M-regular. We apply proposition 7 of the preceding section to $M'=M/\langle\alpha_1,\dots,\alpha_{n-1}\rangle M$ and $M/\alpha M$. There is a $\beta\in\mathfrak{a}$ which is M'- and $M/\alpha M$-active. Therefore $(\alpha_1,\dots,\alpha_{n-1},\beta)$ and (α,β) are M-regular. Then $(\beta,\alpha_1,\dots,\alpha_{n-1})$ and (β,α) are M-regular by the Commuting Lemma, i.e. $(\alpha_1,\dots,\alpha_{n-1})$ is $M/\beta M$-regular and α is $M/\beta M$-active. By induction hypothesis there is an $M/\beta M$-regular sequence $(\alpha,\gamma_2,\dots,\gamma_{n-1})$ in $\mathfrak{a}$. Therefore $(\beta,\alpha,\gamma_2,\dots,\gamma_{n-1})$ is M-regular, and using the Commuting Lemma again, $(\alpha,\beta,\gamma_2,\dots,\gamma_{n-1})$ is M-regular.

The *depth* $d(M)$ of M is the maximal possible length of a M-regular sequence. This depth is always finite because an M-regular sequence $(\alpha_1,\alpha_2,\ldots)$ yields a strictly increasing sequence of ideals $\langle\alpha_1\rangle \subsetneqq \langle\alpha_1,\alpha_2\rangle \subsetneqq \ldots$. We have $d(M)=0$ if and only if all elements of $\mathfrak{m}$ are M-annihilators.

Corollary 4: *If* M *is finitely generated and* α *is* M-*active, then*

$$d(M) = d(M/\alpha M) + 1$$

Proof: The inequality $\geq$ is trivial, the other one $\leq$ follows from Lemma 3.

Let $(\alpha_1,\ldots,\alpha_n)$ be M-regular, let $M_n = M/\langle\alpha_1,\ldots,\alpha_n\rangle M$. The corollary yields by induction on n that $d(M)=d(M_n)+n$. If $(\alpha_1,\ldots,\alpha_n)$ is maximal in the sense, that this sequence cannot be extended to a longer M-regular sequence, there are no M_n-active elements, i.e. $d(M_n)=0$ and hence $d(M)=n$. This implies

Corollary 5: *All maximal* M-*regular sequences have the same length* $d(M)$.

Corollary 6: *If the ring of germs of holomorphic functions in* n *variables* ${}_n\mathcal{O}_o$ *is considered as module over itself,* $d({}_n\mathcal{O}_o) = n$.

Proof: The coordinate functions $z_1,\ldots,z_n$ form a maximal regular sequence.

Lemma 7: *Let* $0 \to A \xrightarrow{f} B \xrightarrow{g} C \to 0$ *be an exact sequence of* R-*modules. If* $\alpha \in \mathfrak{m}$ *is* C-*active, then the induced sequence* $0 \to A/\alpha A \to B/\alpha B \to C/\alpha C \to 0$ *remains exact.*

Proof: The C-activity of α is used only in order to show that $\bar{f}: A/\alpha A \to B/\alpha B$ is injective: Assume that $f(x) \in \alpha B$ i.e. $f(x)=\alpha y$ for some $y \in B$. Then $0=gf(x)=\alpha g(y)$. Since α is C-active, $g(y)=0$. Thus $f(z)=y$ for some $z \in A$. Then $f(\alpha z)=\alpha y=f(x)$. Since f is injective, $x=\alpha z \in \alpha A$.

Lemma 7 implies by induction:

Corollary 8: *If* $(\alpha_1,\dots,\alpha_n)$ *is C-regular, then* $0 \to A/\langle\alpha_1,\dots,\alpha_n\rangle A \to B/\langle\alpha_1,\dots,\alpha_n\rangle B \to C/\langle\alpha_1,\dots,\alpha_n\rangle C \to 0$ *is exact provided* $0 \to A \to B \to C \to 0$ *is exact.*

Proposition 9 (Criterion for Free Modules): *Let* M *be finitely generated, let* $(\alpha_1,\dots,\alpha_n)$ *be* M*-regular and* $\mathfrak{m} = \langle\alpha_1,\dots,\alpha_n\rangle$. *Then* M *is a free* R*-module.*

Proof: Let $k = R/\mathfrak{m}$ denote the residue field. Let the residue classes of $v_1,\dots,v_r \in M$ form a base of the k-vector space $M/\mathfrak{m}M = M \otimes_R k$. Then M is generated by $v_1,\dots,v_r$ (Corollary to Nakayama's lemma on p.81) and so $\pi\colon R^r \twoheadrightarrow M$, $\pi(\xi_1,\dots,\xi_r) = \Sigma\xi_i v_i$ is an epimorphism. Let A = kernel of π. The exact sequence $0 \to A \to R^r \to M \to 0$ induces the exact sequence $0 \to A/\mathfrak{m}A \to R^r/\mathfrak{m}R^r \xrightarrow{\bar\pi} M/\mathfrak{m}M \to 0$ by corollary 8. This is a sequence of k-vector spaces, and $\bar\pi$ is an isomorphism. Therefore $A = \mathfrak{m}A$. Nakayama's lemma implies $A=0$ and so π is an isomorphism.

Observation Concerning the Change of Rings: Let R and S be local noetherian rings, let $\varphi\colon R \to S$ be a local ring homomorphism. Then every S-module M becomes an R-module. The sequence $(\alpha_1,\dots,\alpha_n)$ in R is regular for the R-module M if and only if the image sequence $(\varphi(\alpha_1),\dots,\varphi(\alpha_n))$ in S is regular for the S-module M.

§12 Complete Intersections

Regular sequences and related topics from Commutative Algebra are applied in order to investigate complete intersections locally. As in earlier sections ${}_n\mathcal{O}_o = {}_n\mathcal{O}$ denotes the $\mathbb{C}$-algebra of germs of holomorphic functions in $\mathbb{C}^n$ at 0, see §1, and

$N(\mathfrak{a})$ denotes the zero-set of an ideal $\mathfrak{a} \subset {}_n\mathcal{O}$, see §2.

Lemma 1: *Let M be a finitely generated ${}_n\mathcal{O}$-module. For every prime ideal $\mathfrak{p} \subset {}_n\mathcal{O}$ associated to M we have*

$$\dim N(\mathfrak{p}) \geq \operatorname{depth} M$$

Proof: By induction on $r = \operatorname{depth} M$ beginning with the trivial case $r=0$. Let $\operatorname{depth} M = r \geq 1$ and assume that the lemma holds true for all modules with depth $\leq r-1$. There is an M-active f because $r \geq 1$. Let $P = \{x \in M : \mathfrak{p}x \subset fM\}$; then $fM \subset P$ but

(1) $$fM \neq P .$$

We prove (1) using the submodule $P' = \{x \in M : \mathfrak{p}x = 0\}$. Assume $fM = P$. Then every $x \in P'$ can be written as $x = fx'$, and $0 = \mathfrak{p}x = f\mathfrak{p}x'$. Since f is active, $\mathfrak{p}x' = 0$, i.e. $x' \in P'$. This yields $P' = fP'$ whence by Nakayama's lemma $P' = 0$. But $P' \neq 0$ because $\mathfrak{p}$ is associated to M.

The submodule $U = P/fM$ of M/fM is not zero. Let $\mathfrak{q}$ be a prime ideal which is associated to U. Then $\mathfrak{p} \subset \mathfrak{q}$ because $\mathfrak{p}U = 0$. We have $f \in \mathfrak{q}$ but $f \notin \mathfrak{p}$ because f is active, therefore $\mathfrak{p} \subsetneqq \mathfrak{q}$. Then

(2) $$\dim N(\mathfrak{p}) > N(\mathfrak{q})$$

because of §9, Corollary 4. Since $U \subset M/fM$ the ideal is also associated to M/fM, hence by induction hypothesis

(3) $$\dim N(\mathfrak{q}) > \operatorname{depth} M/fM .$$

Finally we use Corollary 4 of the preceding section

(4) $$\operatorname{depth} M/fM = \operatorname{depth} M - 1$$

and obtain $\dim N(\mathfrak{p}) \geq \operatorname{depth} M$ from (2), (3), and (4).

Proposition 2: *A sequence $(f_1, \ldots, f_n)$ is regular for ${}_n\mathcal{O}$ as module over itself if and only if $N(f_1, \ldots, f_n) = \{0\}$.*

Proof: The Active Lemma 3 in §9 implies one direction by induction on i: For a regular sequence $(f_1, \ldots, f_n)$ we have $\dim N(f_1, \ldots, f_i) = n-i$. Vice versa let $M_i = {}_n\mathcal{O}/\langle f_1, \ldots, f_i\rangle$ and $M_o = {}_n\mathcal{O}$. We must show that f_{i+1} is M_i-active. If this were not so, then $f_{i+1} \in \mathfrak{p}$ for some ideal $\mathfrak{q}$ associated to M_i. Then $\dim N(\mathfrak{q}) \geq \operatorname{depth} M_i = n-i$. Here the inequality follows from lemma 1 above and the equality from the corollaries 4 and 6 of the preceding section. We have $N(\mathfrak{q}) \subset N(f_1,\ldots,f_i)$ because $\mathfrak{q} \supset \langle f_1,\ldots,f_i\rangle$. Obviously $\dim N(f_1,\ldots,f_i) \leq n-i$ because $N(f_1,\ldots,f_n)=\{0\}$. Therefore $\dim N(f_1,\ldots,f_i) = n-i$, and $V=N(\mathfrak{q})$ is an irreducible component of $N(f_1,\ldots,f_i)$. We restrict attention to V. We have $f_1|V=\ldots=f_i|V=0$, and $f_{i+1}|V=0$ because $f_{i+1}\in\mathfrak{q}$. Therefore $0\in N(f_{i+2}|V,\ldots,f_n|V) \subset N(f_1,\ldots,f_n)=\{0\}$, i.e. $\dim V \leq n-i-1$, a contradiction.

Every analytic germ $(Y,0) \subset (\mathbb{C}^n,0)$ is the intersection $Y=X_1\cap\ldots\cap X_r=N(f_1,\ldots,f_r)$ of finitely many hypersurfaces $X_i=N(f_i)$. In general $\dim Y \geq \dim\inf Y \geq n-r$. The germ Y is called a *complete intersection* if and only if there are $r=n-\dim Y$ many hypersurfaces $X_1,\ldots,X_r$ such that $Y = X_1\cap\ldots\cap X_r$.

The ideal $\mathfrak{a}$ is called a complete intersection if $\dim N(\mathfrak{a}) = \delta(\mathfrak{a})$, compare §9(4). Thus X is a complete intersection if and only if $X = N(\mathfrak{a})$ for some complete intersection ideal $\mathfrak{a}$.

Characterization of Complete Intersection Ideals: *The following statements for a proper ideal $\mathfrak{a} \subset {}_n\mathcal{O}$ are equivalent.*

(a) The members of every minimal set of generators of $\mathfrak{a}$ form a regular sequence.

(b) The ideal $\mathfrak{a}$ is generated by the members of a regular sequence.

(c) $\delta(\mathfrak{a}) = \dim N(\mathfrak{a})$ (d) $\delta(\mathfrak{a}) = \dim\inf N(\mathfrak{a})$.

The proof of "(a) implies (b)" is trivial. Proof of "(b) implies (c)": Let $(g_1,\ldots,g_r)$ be a regular sequence such that $\mathfrak{a} = \langle g_1,\ldots,g_r\rangle$. The Active Lemma 3 in §9 implies by induction on r that $\dim N(\mathfrak{a}) = n-r \leq \delta(\mathfrak{a})$. The other inequality as well as "(c) implies (d)" follow from (4) at the end of §9. Proof of "(d) implies (a)": Let $g_1,\ldots,g_r$ be a minimal set of generators of $\mathfrak{a}$, hence $\delta(\mathfrak{a})=n-r$. Let X be an irreducible component of $N(\mathfrak{a})$ having minimal possible dimension $q = \dim X = \dim \inf N(\mathfrak{a})$. There are $f_1,\ldots,f_g \in {}_n\mathcal{O}$ such that $N(f_1|X,\ldots,f_q|X) = \{0\}$, therefore $N(g_1,\ldots,g_r,f_1,\ldots,f_q) = \{0\}$. By (d) we have $r+q=n$. Then $(g_1,\ldots,g_r,f_1,\ldots,f_q)$ is a regular sequence because of proposition 2; in particular the first part $(g_1,\ldots,g_r)$ is a regular sequence.

The statements (c) or (d) show that the property of being a complete intersection ideal $\mathfrak{a}$ depends only on the $\mathbb{C}$-algebra ${}_n\mathcal{O}/\mathfrak{a}$. We observe that regular ideals and principal ideals are complete intersection ideals.

<u>Theorem 3:</u> *Let* $F: (\mathbb{C}^n,0) \to (\mathbb{C}^k,0)$ *be finite and strict with respect to the complete intersection ideal* $\mathfrak{a}$. *Then* ${}_n\mathcal{O}/\mathfrak{a}$ *is a free* ${}_k\mathcal{O}$*-module via* $F^o: {}_k\mathcal{O} \to {}_n\mathcal{O}$.

Proof: We have $\dim N(\mathfrak{a}) = k$ by §9, Propositions 2 and 8. Let $(g_1,\ldots,g_{n-k})$ be a regular sequence such that $\mathfrak{a} = \langle g_1,\ldots,g_{n-k}\rangle$ according to part (b) of the preceding characterization of complete intersections. Let $f_1,\ldots,f_k$ be the components of F . Then $N(g_1,\ldots,g_{n-k},f_1,\ldots,f_k) = \{0\}$ by §6, Theorem 4(a). Therefore the sequence $(g_1,\ldots,g_{n-k},f_1,\ldots,f_k)$ is regular by proposition 2. Then $(f_1,\ldots,f_k)$ is regular for the ${}_n\mathcal{O}$-module ${}_n\mathcal{O}/\mathfrak{a}$. Let $y_1,\ldots,y_k$ denote the coordinate functions in ${}_k\mathcal{O}$. Using the change of rings we conclude that $(y_1,\ldots,y_k)$ is regular for the ${}_k\mathcal{O}$-module ${}_n\mathcal{O}/\mathfrak{a}$. This implies the theorem because of §11, Proposition 9.

If the proof of this theorem is unrolled, almost the whole Chapter III proves to be involved. This theorem will be **an important tool in order to obtain different characterization of the multiplicity, see p.196 below.**

§13 Complex Spaces

Though we are mainly interested in the local aspects of complex analytic geometry the basic notion of the global theory will also be employed.

Let X be a topological Hausdorff space. Let $\mathcal{C}_{X,x}$ denote the $\mathbb{C}$-algebra of germs of continuous complex valued functions at x in X. For every $x \in X$ let a $\mathbb{C}$-subalgebra $\mathcal{O}_{X,x} \subset \mathcal{C}_{X,x}$ be distinguished. Let $\mathcal{O} = \{\mathcal{O}_{X,x}: x \in X\}$. A homeomorphism $h: U \to V$ between an open subset $U \subset X$ and an analytic set V induces isomorphisms $h^x: \mathcal{C}_{V,h(x)} \cong \mathcal{C}_{U,x} = \mathcal{C}_{X,x}$, $f \mapsto f \circ h$, for every $x \in X$. Let $\mathcal{O}_{V,h(x)}$ denote the $\mathbb{C}$-algebra of germs of *holomorphic* functions. The homeomorphism h is called a *holomorphic chart* of $(X,\mathcal{O})$ if $h^x(\mathcal{O}_{V,h(x)}) = \mathcal{O}_{X,x}$ for every $x \in U$. If every point of X lies in the domain of a holomorphic chart, $(X,\mathcal{O})$ is called a *complex space*.

There is a more general notion, where the $\mathcal{O}_{X,x}$ need not be subalgebras of $\mathcal{C}_{X,x}$. Our simpler notion is called 'reduced complex space' in the general theory.

A function $f: X \to \mathbb{C}$ is called holomorphic if its germ at every point x lies in $\mathcal{O}_{X,x}$. A continuous map $h: X \to Y$ between complex spaces is called holomorphic if for every $x \in X$ the induced homomorphism $h^x: \mathcal{C}_{Y,h(x)} \to \mathcal{C}_{X,x}$, $f \mapsto f \circ h$, maps $\mathcal{O}_{Y,h(x)}$ into $\mathcal{O}_{X,x}$. If h has a holomorphic inverse it is called biholomorphic. In this case X and Y are called isomorphic.

Every open subset U of a complex space X is a complex space in an obvious way. A subset $Y \subset X$ is called a *complex subspace* if every point of X has a neighbourhood U such that $Y \cap U = \{x \in U: g_1(x) = \ldots = g_q(x) = 0\}$ for finitely many holomorphic functions $g_1, \ldots, g_q: U \to \mathbb{C}$. A complex subspace is a closed subset. It has the following induced structure of a complex space: For every $y \in Y$ the $\mathbb{C}$-algebra $\mathcal{O}_{Y,y}$ is defined to be the image of $\mathcal{O}_{X,y}$ under the restriction map $\mathcal{C}_{X,y} \to \mathcal{C}_{Y,y}$, $f \mapsto f|Y$. Hence there is an epimorphism $\mathcal{O}_{X,y} \twoheadrightarrow \mathcal{O}_{Y,y}$. Its kernel $\mathcal{J}(Y,y) \subset \mathcal{O}_{X,y}$ is called the ideal of Y at y . By the Nullstellensatz $\mathcal{J}(Y,y)$ is the radical of the ideal $\langle g_1,..,g_q\rangle \subset \mathcal{O}_{X,y}$.

A *complex* n-*dimensional manifold* is a complex space such that the range of every holomorphic chart is an open subset of $\mathbb{C}^n$. One- and two-dimensional complex manifolds are called *complex curves* and *surfaces* respectively. An arbitrary complex space X is called *smooth* or *regular* at the point $x \in X$ if an open neighbourhood of x is a complex manifold.

The complex n-space $\mathbb{C}^n$ is identified with the real 2n-space $\mathbb{R}^{2n}$ by means of $(z_1,..,z_n) = (x_1,y_1,..,x_n,y_n)$ where $z_\nu = x_\nu + iy_\nu$ is the decomposition into real and imaginary parts.

Every complex n-*dimensional manifold is a real* 2n-*dimensional differentiable oriented manifold.*

Proof: For any two holomorphic charts h and h' let $F = h' \circ h^{-1}$ be the coordinate change. It is a biholomorphic map between open subsets of $\mathbb{C}^n$. Let $(f_1,..,f_n)$ denote the holomorphic components with decompositon $f_\nu = u_\nu + iv_\nu$ into real and imaginary parts. Then F is also a diffeomorphism between open subsets of $\mathbb{R}^{2n}$ with real components $(u_1,v_1,..,u_n,v_n)$. The Jacobi déterminant of $(u_1,v_1,..,u_n,v_n)$ equals $|J|^2 > 0$,where J is the complex Jacobi determinant of $(f_1,..,f_n)$. This follows from the Cauchy-Riemann equations $\partial u_\nu/\partial x_\lambda = \partial v_\nu/\partial y_\lambda$, $\partial u_\nu/\partial y_\lambda = -\partial v_\nu/\partial x_\lambda$.

CHAPTER IV

QUOTIENT SINGULARITIES AND THEIR RESOLUTIONS

Let the finite group G act holomorphically on the complex n-dimensional manifold M . Then the orbit space M/G is a complex space. This result, which is due to H. Cartan, is presented in §1-§3. For every point $\xi \in M/G$ the germ $(M/G,\xi)$ is isomorphic to the germ $(\mathbb{C}^n/\Gamma,0)$ for a finite subgroup $\Gamma <$ $GL(n,\mathbb{C})$, and so to the germ (V,0) of the affine orbit variety V of Γ , which has been introduced in II§9. Germs of this type are called *quotient singularities*. They have been characterized and classified by D. Prill. But rather than presenting his general results we restrict attention to the finite subgroups of $SL(2,\mathbb{C})$. Then V is regular at every point $\neq 0$ and definitely singular at 0 .

From the algebraic viewpoint the structure of the singularity (V,0) is the structure of the local algebra $\mathcal{O}_{V,o}$. Geometrically the singularity is concentrated in one point 0 . It becomes better visible by a resolution V: The point 0 is removed from V and replaced by a union E of curves such that $\tilde{V}=(V\setminus\{0\})\cup E$ becomes an everywhere regular surface. The details of this construction (§4-§9) require some patience. But finally (§10-§15) we are able to relate the topology of $E\subset\tilde{V}$ to the results of chapter II, especially to the presentation of the finite subgroups $G < SL(2,\mathbb{C})$ by generators and relations (II§5) and to the 3-manifolds S^3/G (II§4).

The resolution of quotient singularities $\mathbb{C}^2/G$ has a long history with important contributions by H. Jung (1908), DuVal (1934), F. Hirzebruch (1952), and E. Brieskorn (1968). Brieskorn includes all finite subgroups of $GL(2,\mathbb{C})$, not only those of $SL(2,\mathbb{C})$. For another presentation on a more advanced level see the exposé of Pinkham.

§1 Germs of Invariant Holomorphic Functions

Let $G < GL(n,C)$ be a finite subgroup. The elements $\gamma \in G$ act on the holomorphic functions f which are defined in a neighbourhood of 0 in $\mathbb{C}^n$,

$$(f\gamma)(z)=f(\gamma z) .$$

This induces an action of G from the right on the local algebra ${}_n\mathcal{O}_0 = {}_n\mathcal{O}$ of germs of holomorphic functions at 0 in $\mathbb{C}^n$. If $f\gamma=f$ for every $\gamma \in G$, the germ f is called G-invariant. The G-invariant germs form a local subalgebra $\mathcal{O}^G \subset {}_n\mathcal{O}$ with maximal ideal $\mathfrak{m}^G = {}_n\mathfrak{m} \cap \mathcal{O}^G$. If $z \in \mathbb{C}^n$ is not necessarily the origin, the isotropy subgroup $G_z < G$ acts on ${}_n\mathcal{O}_z$. The G_z-invariant germs form the subalgebra $\mathcal{O}^{G_z} \subset {}_n\mathcal{O}_z$.

In II§9 the following results had been obtained: The subalgebra $S^G \subset {}_nS = \mathbb{C}[z_1,\dots,z_n]$ of G-invariant polynomials is generated by finitely many forms $1, f_1,\dots,f_r$ of degree 0 and positive degrees $d_1,\dots,d_r$. The polynomial map

(1) $$F\colon \mathbb{C}^n \to \mathbb{C}^r$$

with components $f_1,\dots,f_r$ induces an epimorphism $F^*\colon {}_rS = \mathbb{C}[x_1,\dots,x_r] \twoheadrightarrow S^G$, $F^*(h)=h\circ F$. The kernel of F^* is a finitely generated ideal $\mathfrak{a} = \langle\varphi_1,\dots,\varphi_s\rangle \subset {}_rS$. It defines the affine orbit variety $V=F(\mathbb{C}^n)=\{x \in \mathbb{C}^r\colon \varphi_1(x)=\dots=\varphi_s(x)=0\}$. The map F is proper and closed. If the range is restricted to V, it is also open. It induces a homeomorphism $F\colon \mathbb{C}^n/G \to V$. Obviously F is holomorphic, and V is an analytic set.

Theorem: *Let* $z \in \mathbb{C}^n$ *with* $F(z)=w \in V \subset \mathbb{C}^r$ *. The image of the induced homomorphism* $F^z: {}_r\mathcal{O}_w \to {}_n\mathcal{O}_z$ *is* $\mathcal{O}^{G_z}$ *. The induced homomorphism* $F^z: \mathcal{O}_{V,w} \to {}_n\mathcal{O}_z$ *maps* $\mathcal{O}_{V,w}$ *isomorphically onto* $\mathcal{O}^{G_z}$ *.*

The second statement follows from the first one because $F^z: {}_r\mathcal{O}_w \to \mathcal{O}_{V,w} \xrightarrow{F^z} {}_n\mathcal{O}_z$ is an epimorphism followed by a monomorphism (because $F: \mathbb{C}^n \to V$ is open). The inclusion $F^z({}_r\mathcal{O}_w) \subset \mathcal{O}^{G_z}$ is obvious. But the proof of the other inclusion

$$\mathcal{O}^{G_z} \subset F^z({}_r\mathcal{O}_w) \tag{2}$$

is rather long. First the point $z=0$ is considered. Then $w=0$ and $G_z=G$. For short the indices z and w are omitted in this case. We must prove

$$\mathcal{O}^G \subset F^o({}_r\mathcal{O}) \quad , \tag{3}$$

and begin with

Lemma 1: *If* ${}_n\mathcal{O}$ *is considered as* ${}_r\mathcal{O}$*-module by means of* $F^o: {}_r\mathcal{O} \to {}_n\mathcal{O}$ *, it is finitely generated.*

Proof: According to the Preparation Theorem (III§5) it suffices to show that ${}_n\mathcal{O}/{}_r\mathfrak{m}\cdot{}_n\mathcal{O} = {}_n\mathcal{O}/\langle f_1,\dots,f_r\rangle$ is a finite dimensional complex vector space. By proposition 2 on p.100 this is equivalent to $N(f_1,\dots,f_r) = \{0\}$. Now $N(f_1,\dots,f_r) = F^{-1}(0)$, and $F^{-1}(0)=\{0\}$ by (7) on p.66.

Methods from Local Algebra will now yield the proof of (3): If $\mathfrak{m}$ is an ideal of the ring R , the $\mathfrak{m}$-topology on R is defined by means of the basis of neighbourhoods of 0 consisting of the powers $\mathfrak{m}^k$, $k=0,1,\dots$. Thus R becomes a topological ring. More generally let A be an R-module.

The $\mathfrak{m}$-topology on A is defined by means of the basis of neighbourhoods of 0 consisting of the $\mathfrak{m}^k A$, $k=0,1,\dots$. This topology is separated (i.e. A is a Hausdorff space) if and only if $\{0\}$ is closed in A , equivalently

$$\bigcap_{k=0}^{\infty} \mathfrak{m}^k = \{0\} .$$

Lemma 2: *Let* R *be a local noetherian ring with maximal ideal* $\mathfrak{m}$ *. Every submodule* B *of a finitely generated* R-*module* A *is closed in the* $\mathfrak{m}$-*topology, in particular* (B=0) : *The* $\mathfrak{m}$-*topology is separated.*

Proof: First the special case $B=\{0\}$ is considered. Since R is noetherian and A is finitely generated, $C = \bigcap_{k=0}^{\infty} \mathfrak{m}^k A$ is also finitely generated. Obviously $\mathfrak{m}C=C$. Nakayama's lemma implies $C=\{0\}$. For the general case we observe that the projection $p: A \twoheadrightarrow A/B$ is continous (as any linear mapping of R-modules). Since A/B is finitely generated, $\{0\} \subset A/B$ is closed and hence $B=p^{-1}(0)$ is closed in A .

Lemma 3: *Let* k *be a field,* R *a local noetherian* k-*algebra with maximal ideal* $\mathfrak{m}$ *. Let* $\mathfrak{n}$ *be another ideal so that* $R/\mathfrak{n}$ *is a finite dimensional* k-*vector space. Then the* $\mathfrak{m}$- *and the* $\mathfrak{n}$-*topology on* R *coincide.*

Proof: Since $\mathfrak{n} \subset \mathfrak{m}$, the $\mathfrak{n}$-topology is finer than the $\mathfrak{m}$-topology. In order to prove the converse we must show that $\mathfrak{m}^r \subset \mathfrak{n}$ for some r . But this is equivalent to $\dim_k R/\mathfrak{n} < \infty$ by the proof of proposition 2 on p.100.

We shall apply these results to the local noetherian ring $\mathcal{O} = {}_n\mathcal{O}$ of germs of holomorphic functions with its maximal ideal $\mathfrak{m}$. The $\mathfrak{m}$-topology on $\mathcal{O}$ is separated because of lemma 2. The polynomial ring $S \subset \mathcal{O}$ is dense. Every element $\gamma \in GL(n,\mathbb{C})$ acts continously on $\mathcal{O}$ because of $\mathfrak{m}\gamma = \mathfrak{m}$. Therefore the mapping $\varphi_\gamma : \mathcal{O} \to \mathcal{O}$, $f \mapsto f - f\gamma$, is continous. Since $\mathcal{O}$ is separated, the inverse image $\{f: f\gamma = f\} = \varphi_\gamma^{-1}(0)$ is closed in

$\mathcal{O}$. Hence for any subgroup $G < GL(n,\mathbb{C})$ the subring of invariant germs $\mathcal{O}^G = \bigcap_{\gamma\in G} \varphi_\gamma^{-1}(0)$ is closed in $\mathcal{O}$. Since $S^G = S \cap \mathcal{O}^G$, the *closure of* S^G *in* $\mathcal{O}$ *is* $\mathcal{O}^G$.

The ${}_r\mathcal{O}$-submodule $F^o({}_r\mathcal{O}) \subset {}_n\mathcal{O}$ is closed with respect to the ${}_r\mathfrak{m}$-topology on the ${}_r\mathfrak{m}$-module ${}_n\mathcal{O}$ because of lemma 1 and 2. Let us consider the ideal $\mathfrak{n} = {}_r\mathfrak{m}\,{}_n\mathcal{O}$ of the ring ${}_n\mathcal{O}$. We have $\mathfrak{n}^k = {}_r\mathfrak{m}^k \cdot {}_n\mathcal{O}$ so that the ${}_r\mathfrak{m}$-topology on the ${}_r\mathcal{O}$-module ${}_n\mathcal{O}$ and the $\mathfrak{n}$-topology on the ring ${}_n\mathcal{O}$ coincide. By lemma 1 ${}_n\mathcal{O}/\mathfrak{n}$ is a finite dimensional $\mathbb{C}$-vector space. Therefore lemma 3 implies that the ${}_n\mathfrak{m}$-topology and the $\mathfrak{n}$-topology on ${}_n\mathcal{O}$ coincide. So $F^o({}_r\mathcal{O}) \subset {}_n\mathcal{O}$ is also closed with respect to the ${}_n\mathfrak{m}$-topology. We know $S^G = F^o({}_rS)$ from II§9. The closure of S^G in ${}_n\mathcal{O}$ with respect to the ${}_n\mathfrak{m}$-topology is $\mathcal{O}^G$. Since $F^o({}_rS) \subset F^o({}_r\mathcal{O})$ and $F^o({}_r\mathcal{O})$ is closed, the inclusion (3) $\mathcal{O}^G \subset F^o({}_r\mathcal{O})$ follows.

In order to obtain the more general result (2) for arbitrary points $z \in \mathbb{C}^n$ we need conditions for a holomorphic germ

$$H: (\mathbb{C}^n,0) \to (\mathbb{C}^r,0)$$

such that it behaves like the germ of F at 0. Let $d_j = \deg f_j$.

<u>Proposition 4</u>: *If the components* h_j *of* H *belong to* $\mathcal{O}^G$ *and if* $h_j - f_j \in \mathfrak{m}^{d_j+1}$, *then* $H^o: {}_r\mathcal{O} \to \mathcal{O}^G$ *is epimorphic.*
For the proof it suffices to find an invertible holomorphic germ $\Phi: (\mathbb{C}^r,0) \to (\mathbb{C}^r,0)$ with $H = \Phi \circ F$. We order the components $f_1,\dots,f_r$ of F such that the first q ones $(q \le r)$ form a minimal set of generators of S^G and such that $d_1 \le \dots \le d_q$. Let $F_1: \mathbb{C}^n \to \mathbb{C}^q$ denote the map with the first q components $f_1,\dots,f_q$. It has been proved above, see (3), that $F_1^o: {}_q\mathcal{O} \twoheadrightarrow \mathcal{O}^G$ is epimorphic. Hence there are $\psi_1,\dots,\psi_r$

$\in {}_q\mathcal{O}$ such that $h_j=\psi_j\circ F_1$ for $1\leq j\leq q$ and $h_j-f_j=\psi_j\circ F_1$ for $q+1\leq j\leq r$. The components of Φ are defined to be $\varphi_j(x_1,\ldots,x_r) =\psi_j(x_1,\ldots,x_q)$ for $1\leq j\leq q$ and $\varphi_j(x_1,\ldots,x_r)=x_j+\psi_j(x_1,\ldots,x_q)$ for $q+1\leq j\leq r$. Then $H=\Phi\circ F$ is easily checked. In order to show that Φ is invertible, the Jacobian matrix $D\Phi$ at 0 is calculated. The following result will be proved below:

Lemma 5: *Let the components* f_j *of* $F_1\colon \mathbb{C}^n \to \mathbb{C}^q$ *be forms of degree* d_j, *which form a minimal set of generators of* S^G. *Let* $d_1\leq\ldots\leq d_q$. *For every* $h\in {}_q\mathcal{O}$ *with* $h\circ F_1\in {}_n\mathfrak{m}^{d_j+1}$ *the partial derivatives at the origin* $\partial h/\partial x_i=0$ *for* $i<j$.

For $1\leq j\leq q$ we have $(\psi_j-x_j)\circ F=h_j-f_j\in \mathfrak{m}^{d_j+1}$, therefore by the lemma $\partial\psi_j/\partial x_i=\delta_{ij}$ (Kronecker symbol) for $i\leq j$. This implies $\partial\varphi_j/\partial x_j=\delta_{ij}$ for $1\leq i\leq j\leq q$. Further, $\partial\varphi_j/\partial x_i=0$ for $1\leq j\leq q$ and $q+1\leq i\leq r$. For $q+1\leq i,j\leq r$ we have $\partial\varphi_j/\partial x_i=\delta_{ij}$. Thus the Jacobian $D\Phi$ at 0 has the form

$$\left[\begin{array}{ccc:ccc} 1 & & * & & & \\ & \ddots & & & 0 & \\ 0 & & 1 & & & \\ \hdashline & & & 1 & & 0 \\ & * & & & \ddots & \\ & & & 0 & & 1 \end{array}\right],$$

which proves the invertibility.

In order to prove Lemma 5 *weighted homogeneous polynomials* will be introduced. Let $d_1,\ldots,d_r$ be natural numbers. A polynomial $p(x_1,\ldots,x_r)$ is called weighted homogeneous with respect to $(d_1,\ldots,d_r)$ of weight d if $p(t^{d_1}x_1,\ldots,t^{d_r}x_r) = t^d p(x_1,\ldots,x_r)$. Every $h\in {}_r\mathcal{O}$ can uniquely be written as convergent series $h=\sum_{d=0}^{\infty} h_d$ of weighted homogeneous polynomials of weight d for a fixed set $(d_1,\ldots,d_r)$.

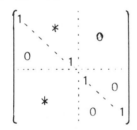

Let $f_1,\ldots,f_r$ be forms in $z_1,\ldots,z_n$ of degrees $d_1,\ldots,d_r$. Let $F\colon \mathbb{C}^n\to\mathbb{C}^r$ be the polynomial map with components f_i. If $p(x_1,\ldots,x_r)$ is weighted homogeneous of weight d, then $p\circ F = p(f_1,\ldots,f_r)$ is a form of degree d. If $h=\Sigma h_d$ is a convergent series of weighted homogeneous polynomials of weight d, then $h \circ F = \Sigma h_d \circ F$ is the usual Taylor series. The proof of lemma 5 begins with the following

Lemma 6: *Under the hypotheses of lemma 5 let* h *be weighted homogeneous with respect to* $(d_1,\dots,d_q)$. *If* $h \circ F_1 = 0$, *then* $h \in {}_q\mathfrak{m}^2$.

Proof: A weighted homogeneous $h \notin {}_q\mathfrak{m}$ is constant, hence $h=0$ because $h \circ F_1 = 0$. If $h \in {}_q\mathfrak{m} \setminus {}_q\mathfrak{m}^2$ at least one variable x_k occurs linearly in h . Therefore $h = ax_k + h^*$ with $a \neq 0$ and with no linear term x_k in h^* . Since h is weighted homogeneous, h^* is also weighted homogeneous and all three, h, x_k, and h^* have the same weight d_k . Then x_k cannot occur in h^* . The hypothesis $h \circ F_1 = 0$ implies $-af_k = h^*(f_1,\dots,\hat{f}_k,\dots,f_r)$, i.e. f_k is superfluous as generator of S^G .

Proof of lemma 5: We write $h = \Sigma h_d$ as series of weighted homogeneous polynomials, so that $h \circ F_1 = \Sigma h_d \circ F_1$ is the Taylor series. The assumption $h \circ F_1 \in \mathfrak{m}^{d_j+1}$ implies $h_d \circ F_1 = 0$ for $d \leq d_j$, therefore $h_d \notin {}_r\mathfrak{m}^2$ for $d \leq d_j$ according to lemma 6. Assume now $\partial h / \partial x_j \neq 0$. Then x_j occurs linearly in h , namely in the term h_{d_j} . But h_{d_j} has no linear term unless $d_i > d_j$, i.e. $i > j$.

This concludes the proof of proposition 4, which will be the main tool in order to obtain the inclusion (2) for $z \neq 0$. Furthermore we need:

Lemma 7: *Given a positive* d *and a* G_z*-invariant polynomial* p . *There is a* G*-invariant polynomial* q *so that* p-q *has order* >d *at* z , *i.e.* $p - q \in \mathfrak{m}_z^{d+1}$.

Proof: There is a polynomial u which has order >d at every point $x \in G \cdot z$ if $x \neq z$ and for which 1-u has order >d at z . The G_z-invariant polynomial $v = \prod_{\gamma \in G_z} u\gamma$ has the same properties, and so pv has order >d at every $x \in G \cdot z$ if $x \neq z$ and $(v-1)p$ has order >d at z . Let $Q \subset G$ be a set of representatives for the residue classes of G modulo

G_z with 1 representing G_z. Then $q = \sum_{\tau \in Q} (pv)\tau$ has the desired properties.

Proof of (2), $\mathcal{O}_z^{G_z} \subset F^z({}_r\mathcal{O}_w)$: There is a polynomial map $\tilde{F}: \mathbb{C}^n \to \mathbb{C}^s$, with components $\tilde{f}_1, \dots, \tilde{f}_s$ such that ${}_sS \to S^{G_z}$, $\varphi \to \varphi \circ \tilde{F}$ is epimorphic. Let $T: \mathbb{C}^n \to \mathbb{C}^n$ be the translation given by $T(z)=0$. According to lemma 7 there are G-invariant polynomials $h_1, \dots, h_s$ such that $\tilde{f}_j \circ T - h_j$ has "high order" at z. Let $H: \mathbb{C}^n \to \mathbb{C}^s$ be the polynomial map with coefficients h_j. We apply proposition 4 to $H \circ T^{-1}$ and $\mathcal{O}_o^{G_z}$: Since $\tilde{f}_j - h_j \circ T^{-1}$ has high order at 0, the induced homomorphism $(H \circ T^{-1})^o: {}_s\mathcal{O}_o \to \mathcal{O}_o^{G_z}$ is surjective and thus ${}_s\mathcal{O}_o \to \mathcal{O}_z^{G_z}$, $\psi \to \psi \circ H$, is an epimorphism. Since H is G-invariant, there is a polynomial map Φ such that $H = \Phi \circ F$. The composition $H^z: {}_s\mathcal{O}_o \xrightarrow{\Phi^w} {}_r\mathcal{O}_w \xrightarrow{F^z} \mathcal{O}_z^{G_z}$ is surjective, therefore F^z is surjective.

§2 Complex Orbit Spaces

Orbit spaces X/G of group actions can be rather pathological. In order to exclude them one needs extra hypotheses on the action or one restricts attention to actions of *finite* groups G, as we do. Let $p: X \to X/G$ denote the projection.

Proposition 1: *Let the finite group* G *act on the Hausdorff space* X. *Then the orbit space* X/G *is also a Hausdorff space.*

Proof: Let $c, c' \in X$ be such that $p(c) \neq p(c')$, i.e. such that $G\cdot c \cap G\cdot c' = \emptyset$ for the orbits. Since X is a Hausdorff space and the orbits are finite sets, there are disjoint neighbourhoods V of $G\cdot c$ and V' of $G\cdot c'$. By continuity of the action, for every $g\in G$ there is a neighbourhood U_g of c with $gU\subset V$ for every $g\in G$. The finite intersection $U=\bigcap_{g\in G} U_g$ is a neighbourhood of c with $gU\subset V$ for all $g\in G$. Similarly there is a neighbourhood U' of c' with $gU'\subset V'$ for all $g\in G$. Since p is open, $p(U)$ and $p(U')$ are neighbourhoods of $p(c)$ and $p(c')$, respectively. They are disjoint because $V\cap V'=\emptyset$.

Proposition 2: *Let the finite group* G *act on the Hausdorff space* X. *Every* $c\in X$ *has an arbitrary small* G_c*-invariant neighbourhood* U *such that the canonical map* $U/G_c\to p(U)$ *is a homeomorphism. Here* $G_c=\{g\in G: gc=c\}$ *denotes the isotropy subgroup of* c.

Proof: It suffices to find a G_c-invariant U such that for any $x,y\in U$ with $p(x)=p(y)$ there is a $g\in G_c$ with $y=g_c x$. Since the orbit $G\cdot c$ is finite, there are arbitrary small mutually disjoint neighbourhoods V_z of the points $z\in G\cdot z$. There is a neighbourhood W of c such that $gW\subset V_{gc}$ for every $g\in G$. Then $U=\bigcup_{g\in G_c} gW$ is a G_c-invariant neighbourhood with the desired property.

Let X be a complex n-dimensional manifold, compare III§13. Let the finite group G act holomorphically on X. For every point $\xi\in X/G$ of the orbit space let $\mathcal{C}_{X/G,\xi}$ denote the $\mathbb{C}$-algebra of germs of continuous complex valued functions at ξ. The following subalgebra $\mathcal{O}_{X/G,\xi}$ is distinguished:

(1) *A germ* $f\in\mathcal{C}_{X/G,\xi}$ *belongs to* $\mathcal{O}_{X/G,\xi}$ *if for every* $x\in X$ *with* $p(x)=\xi$ *the germ* $f\circ p\in\mathcal{O}_{X,x}$.

It suffices to require $f\circ p\in\mathcal{O}_{X,x}$ for *one* x: If also $p(y)=\xi$, there is a $g\in G$ with $x=gy$. Since g acts holomorphically every germ $h\in\mathcal{O}_{X,x}$ is transformed into $hg\in\mathcal{O}_{X,y}$, in particular $f\circ p\in\mathcal{O}_{X,x}$ is transformed into $f\circ p\in\mathcal{O}_{X,y}$.

Since p is open, it induces a *monomorphism* $p^x\colon \mathcal{O}_{X/G,\xi} \to \mathcal{O}_{X,x}$, $f \mapsto f\circ p$. The image of p^x is the subalgebra $\mathcal{O}_{X,x}^{G_x}$ of G_x-invariant holomorphic germs. This follows from proposition 2, thus

(2) $$p^x\colon \mathcal{O}_{X/G,\xi} \cong \mathcal{O}_{X,x}^{G_x} \quad \text{for every} \quad x \in p^{-1}(\xi)\ .$$

Lemma 3: *For every* $c \in X$ *there is a holomorphic chart* $h\colon U \to W \subset \mathbb{C}^n$ *such that* $h(c)=0$ *and such that* G_c *acts linearly on* W *by means of* hgh^{-1} *for every* $g \in G_c$.

Proof: We begin with an arbitrary holomorphic chart $h^*\colon U^* \to W^*$ with a G_c-invariant domain U^* so that $h^*(c)=0$. For every $g \in G_c$ we have the holomorphic map $g^* = h^*gh^{*-1}$ defined on W^*. Let $g' = D_0g^* \in GL(n,\mathbb{C})$ denote the differential of g^* at 0, and define $\varphi(z) = (\#G_c)^{-1} \sum_{g\in G_c} g'^{-1}g^*(z)$ for $z \in W^*$. Then $D_0\varphi = \mathrm{id}$ so that φ maps a (eventually smaller) W^* biholomorphically onto a neighbourhood W of 0 in $\mathbb{C}^n$. We replace the chart h^* by $h=\varphi h^*$. Then $hgh^{-1}=g'$ acts linearly.

Theorem 4: *Let the finite group* G *act holomorphically on the complex* n*-dimensional manifold* X . *The orbit space* X/G *with the complex structure* $\mathcal{O}$ *defined by* (1) *is a complex space: For every point* $c \in X$ *there is a representation (homomorphism)* $\rho_c\colon G_c \to GL(n,\mathbb{C})$ *and a holomorphic chart* $h^*\colon U^* \to V^*$ *whose domain* U^* *is a neighbourhood of* $p(c) \in X/G$ *and whose range* V^* *is a neighbourhood of* 0 *in the affine orbit variety of the linear action of* $\rho_c(G_c)$ *on* $\mathbb{C}^n$.

Proof: Using lemma 3 the representation is defined to be $\rho_c(g)=hgh^{-1}$. Let $F\colon \mathbb{C}^n \to V$ denote the polynomial map onto the affine orbit variety V of the linear action of $\Gamma_c = \rho_c(G_c)$, see II§9. For W as in lemma 3 the image $V^*=F(W)$ is a neighbourhood of 0 in V because F is open, see II§9(9). We consider the commutative diagram 1. By proposition 2 and lemma 3 the chart h induces the homeomorphism $\bar{h}$. The map F induces the homeomorphism $\bar{F}\colon \mathbb{C}^n/\Gamma_c \to V$, see

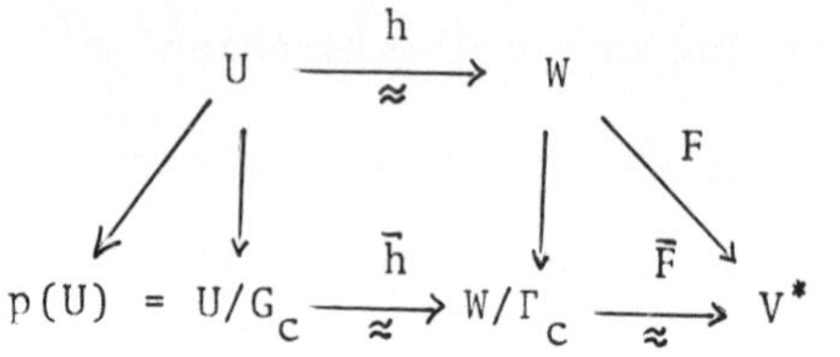

Diagram 1

II§9(10), which maps W/Γ_c onto V^* . Then $h^*=\bar{F}\circ\bar{h}$: $U^*=p(U)\to V^*$ is a holomorphic chart: Let $x\in U$, $y=Fh(x)\in V^*$ and $\xi=p(x)\in U^*$. If $f\in \mathcal{O}_{V,y}$, then $fh^*\in \mathcal{O}_{X/G,\xi}$ because $fh^*p=fFh\in \mathcal{O}_{X,x}$. Vice versa, if $\varphi\in\mathcal{O}_{X/G,\xi}$ is given, we find an $f\in\mathcal{O}_{V,y}$ with $\varphi=fh^*$ as follows. We have $\varphi p\in \mathcal{O}^{G_x}_{X,x}$ by (2) , therefore $\varphi ph^{-1}\in {}_n\mathcal{O}^{\Gamma_x}_x$, where $\Gamma_x<\Gamma_c$ denotes the isotropy subgroup of $h(x)$. By the main result (2) of §1 there is an $f\in\mathcal{O}_{V,y}$ with $\varphi ph^{-1}=fF$. This implies $\varphi p=fh^*p$, hence $\varphi=fh^*$ because p is open.

§3 Quotient Singularities

The germ (X,x) of an analytic set or equivalently of a complex space is called a *singularity*, though it may be regular. According to III§3, Corollary 2, (X,x) is determined up to isomorphy by the $\mathbb{C}$-algebra $\mathcal{O}_{X,x}$ of the germs of holomorphic functions. The singularity (X,x) is irreducible if $\mathcal{O}_{X,x}$ is entire, see III§2, Corollary. It is called *normal* if in addition $\mathcal{O}_{X,x}$ is integrally closed. For the definition of "integrally closed" see the Algebraic Preliminaries in III§7. A regular (X,x) is normal because then $\mathcal{O}_{X,x}\cong {}_n\mathcal{O}$, and ${}_n\mathcal{O}$ is integrally closed being a factorial ring. The cone $z^2=xy$ at the origin is a non-regular normal singularity. The plane cusp $z^2=w^3$ at the origin is an irreducible non-normal singularity. Normal singularities are intermediate between arbitrary singularities and regular ones. Their properties and the "normalization" of arbitrary singularities are extensively studied in Chapter 6,7, and 8 of Grauert-Remmert.

Proposition 1: *Let* V *be the affine quotient of a finite subgroup* $G < GL(n,\mathbb{C})$. *The singularity* $(V,0)$ *is irreducible, normal,* n-*dimensional.*

Proof: According to the theorem in §1 $\mathcal{O}_{V,o} \cong {}_n\mathcal{O}^G$. Since ${}_n\mathcal{O}^G$ is a subring of the entire ring ${}_n\mathcal{O}$, it is also entire, hence $(V,0)$ is irreducible. If a finite group acts on an entire integrally closed ring, the subring of G-invariant elements is obviously integrally closed, too. Therefore ${}_n\mathcal{O}^G$ is integrally closed and so $(V,0)$ is normal. Let $q=\dim_o V$. By III§9, proposition 8, and III§6, Theorem 8, there is a finite and strict map $H: (V,0) \to (\mathbb{C}^q,0)$. The polynomial map $F: (\mathbb{C}^n,0) \to (V,0)$ is open and has finite fibres. Therefore $H\circ F: (\mathbb{C}^n,0) \to (\mathbb{C}^q,0)$ is finite and strict (III§8 Theorem 5). This implies $n=q$.

A singularity (X,x) is called a *quotient singularity* if (X,x) is isomorphic to the germ $(V,0)$ of the affine orbit variety of a finite subgroup $G < GL(n,\mathbb{C})$. By §2 the complex quotient space M/G of the action of a finite group G on the complex n-dimensional manifold M consists of n-dimensional quotient singularities only.

Proposition 2: *Every one-dimensional quotient singularity is regular.*

Proof: For every $m=1,2,\dots$ there is exactly one finite subgroup $G_m < GL(1,\mathbb{C}) = \mathbb{C}^\times$, namely the group of m-th roots of unity. The map $\mathbb{C} \to \mathbb{C}$, $z \mapsto z^m$, induces an isomorphism $\mathbb{C}/G_m \cong \mathbb{C}$.

Proposition 3: *Let* $G < SL(2,\mathbb{C})$ *be a finite subgroup.*

(a) *For every* $z \in \mathbb{C}^2$, $z \neq 0$, *the isotropy subgroup* $G_z=\{id\}$.

(b) *The polynomial map* $F: \mathbb{C}^2 \to V$ *onto the orbit variety is locally biholomorphic at every point* $z \neq 0$.

(c) *The orbit variety* V *is regular at all points* $\neq 0$.

Proof: Let the coordinates in $\mathbb{C}^2$ be chosen such that z is the first base vector. Then every $A \in G_z$ has the form $A = \begin{pmatrix} 1 & b \\ 0 & c \end{pmatrix}$ with $c=1$ because $\det A=1$. We have $A^n = \begin{pmatrix} 1 & nb \\ 0 & 1 \end{pmatrix}$. Since $A^n=E$ for some $n<\infty$, $b=0$. This proves (a). Part (b) follows from (a) because F^z maps $\mathcal{O}_{V,w}$ isomorphically onto $\mathcal{O}^{G_z}$ for $w=F(z)$. Finally (c) follows immediately from (b).

The orbit variety V of a finite subgroup $G < SL(2,\mathbb{C})$ is a surface in $\mathbb{C}^3$ given by one equation $\varphi=0$, see table 4 in II§8. The defining polynomials φ are weighted homogeneous with respect to (d_1,d_2,d_3) of weight d according to the following table.

Type of the group	φ	$d_1,d_2,d_3;\ d$
C_n	$x^2-y^2+z^n$	$n,\ n,\ 2;\ 2n$
$\tilde{D}_q$	$x^2+y^2z-z^{q+1}$	$q+1,\ q,\ 2;\ 2(q+1)$
$\tilde{T}$	$x^2+4y^3-z^4$	$6,\ 4,\ 3;\ 12$
$\tilde{O}$	$x^2+y^3z+z^3$	$9,\ 4,\ 6;\ 18$
$\tilde{I}$	$x^2+y^3+z^5$	$15,10,\ 6;\ 30$

The polynomials φ are irreducible elements of ${}_3\mathcal{O}$. Therefore the principal ideal $\langle\varphi\rangle = \mathfrak{J}(V)$ is the ideal of the germ $(V,0) \subset (\mathbb{C}^3,0)$. Obviously $\langle\varphi\rangle$ is minimally embedded, see III§4. Therefore $(V,0) \subset (\mathbb{C}^3,0)$ is minimally embedded, in particular, V is not regular at 0.

If $G < GL(2,\mathbb{C})$ is finite, but not contained in $SL(2,\mathbb{C})$, it is still true, that the orbit variety V is regular at every point $\neq 0$. It may even occur that V is also regular at 0. Using proposition 2 examples are easily found: Let η be a primitive n-th root of unity, $n \geq 2$. Let G be the cyclic group generated by $\begin{pmatrix} 1 & 0 \\ 0 & \eta \end{pmatrix}$. Then $\mathbb{C}^2 \to \mathbb{C}^2$, $(z,w) \mapsto (z,w^n)$ induces an isomorphism $\mathbb{C}^2/G \cong \mathbb{C}^2$. On the other hand there are finite subgroups G of $GL(2,\mathbb{C})$, such that the quotient singularity $\mathbb{C}^2/G$ is not a surface in $\mathbb{C}^3$ but has an embedding dimension ≥ 4: Take e.g. the cyclic group of order 3 generated by $\begin{pmatrix}\eta & 0\\ 0 & \eta\end{pmatrix}$, where η is a third root of unity , $\eta \neq 1$. The subalgebra of invariant polynomials $S^G \subset \mathbb{C}[z,w]$ is generated by $x=z^3$, $y=z^2w$, $u=zw^2$, $v=w^3$. There are three relations $xv=yu$, $xu=y^2$, $yv=u^2$. The polynomial map $F: \mathbb{C}^2 \to \mathbb{C}^4$ with components z^3, z^2w, zw^2, w^3 maps $\mathbb{C}^2$ onto $V=F(\mathbb{C}^2)=\{(x,y,u,v)\in\mathbb{C}^4: xv=yu,\ xu=y^2,\ yv=u^2\}$. The ideal of $(V,0)$ in ${}_4\mathcal{O}$ is $\mathcal{J}(V) = \langle xv-yu,\ xu-y^2,\ yv-u^2\rangle$. It is minimally embedded and requires three generators. Therefore $\mathrm{edim}_0 V=4$, and V is not a complete intersection at 0 .

If V is the affine quotient variety of a finite group $G < SL(2,\mathbb{C})$, then V **is isomorphic to the zero set** $N(\varphi)$ of a polynomial φ of the table above. As far as the uniqueness of the germ $\varphi \in {}_3\mathcal{O}$ is concerned, there is the following

Proposition 4: *Let* $f \in {}_3\mathcal{O}$ *be a germ without multiple factors. If* $N(f)$ *is isomorphic to the quotient singularity* $\mathbb{C}^2/G$ *of a finite subgroup* $G < SL(2,\mathbb{C})$, *there is a biholomorphic germ* $\Phi: (\mathbb{C}^3,0) \cong (\mathbb{C}^3,0)$ *such that* $\varphi = f\circ\Phi$ *is the polynomial which belongs to the type of* G *according to the table above.*

"f has no multiple factors" means: If $f=f_1^{r_1}\cdots f_q^{r_q}$ is the unique factorization into different irreducible f_ν , then $r_1=\ldots=r_q=1$. This assumption is quite natural because for arbitrary r_ν the zero sets $N(f_1^{r_1}\cdots f_q^{r_q})=N(f_1\cdots f_q)$ are equal.

Proof: Let $f\in{}_n\mathcal{O}$ have no multiple factors. Then $\langle f\rangle=\mathrm{rad}\langle f\rangle=\mathcal{J}N(f)$ by III§8. Let $\varphi \in {}_n\mathfrak{m}^2$ also have no multiple factors and assume $N(\varphi) \cong N(f)$. Then $f \in {}_n\mathfrak{m}^2$ and ${}_n\mathcal{O}/\langle f\rangle \cong {}_n\mathcal{O}/\langle\varphi\rangle$. By III§4, proposition, there is a biholomorphic $H: (\mathbb{C}^n,0) \to (\mathbb{C}^n,0)$ such that $\langle f\circ H\rangle=\langle\varphi\rangle$, therefore $f\circ H=u\varphi$ for some unit u . Assume now that φ is weighted homogeneous with respect to $(d_1,\ldots,d_n)$ of weight d . There is a unit v with $v^d=u$. Then $f\circ H=v^d\varphi=\varphi\circ\Psi$ for $\Psi(x_1,\ldots,x_n)=(v^{d_1}x_1,\ldots,v^{d_n}x_n)$. This Ψ is biholomorphic so that $\varphi=f\circ H\circ\Psi^{-1}$ yields the desired result for $n=3$.

§4 Modifications. Line Bundles

A holomorphic mapping $f: Y \to X$ between two complex spaces is called a *modification* of X at the point $a \in X$ with exceptional set $E=f^{-1}(a)$ if

(1) f is proper.

(2) $Y \setminus E$ is dense in Y

(3) $Y \setminus E$ is mapped biholomorphically onto $X \setminus \{a\}$.

The modification f is called a *resolution* of the singularity (X,a) if Y is regular at every point $y \in E$.

Modifications of analytic sets can be inserted into complex spaces and yield all possible modifications of complex spaces: In order to modify X at a we take a holomorphic chart $h: U \to V$ defined in a neighbourhood of a . Let $\varphi: \tilde{V} \to V$ be a modification of V at $h(a)$. In the disjoint union of $\tilde{V}$ and $X \setminus \{a\}$ we identify the point $y \in \tilde{V}$ with $x \in X \setminus \{a\}$ if and only if $\varphi(y)=h(x)$. Let $\tilde{X}$ be the identification space. The holomorphic mapping $f: \tilde{X} \to X$ is defined by $f(x)=x$ if $x \in X \setminus \{a\} \subset \tilde{X}$ and by $f(y)=h^{-1}\varphi(y)$ if $y \in \tilde{V} \subset \tilde{X}$.

Using *line bundles* $\Sigma(b)$ over the projective line we shall describe resolutions in some simple cases. Let $b=0,1,\ldots$. The following smooth complex surface is introduced

$$\Sigma(b) = \{(z,w) \in \mathbb{C}^2 \times \mathbb{P}_1 : z_1 w_2^b = z_2 w_1^b \} .$$

Here $z = (z_1, z_2)$ are the coordinates of $\mathbb{C}^2$ and $w=[w_1:w_2]$ are the homogeneous coordinates of the projective line $\mathbb{P}_1$. Obviously $\Sigma(0) = \mathbb{C} \times \mathbb{P}_1$. For $b=1$ the projection

$$\sigma: \Sigma(1) \to \mathbb{C}^2 , \quad \sigma(z,w) = z ,$$

is a modification at the origin with exceptional set $E = \sigma^{-1}(0) = \{(0,w) \in \Sigma(1)\}$ isomorphic to $\mathbb{P}_1$. Since the point 0 is replaced by $E \approx$ real 2-sphere, the step from $\mathbb{C}^2$ to $\Sigma(1)$ is also called *blowing up of the origin*. The mapping σ is called *blowing down of* E.

For $b>1$ the projection $\Sigma(b) \to \mathbb{C}^2$, $\sigma(z,w)=z$, is no longer a modification because for every point $z \neq 0$ the inverse image consists of b different points. But let us consider the cyclic group $C_b < GL(2,\mathbb{C})$ generated by $A = \begin{pmatrix} \eta & 0 \\ 0 & \eta \end{pmatrix}$ with $\eta=e^{2\pi i/b}$ and its orbit projection $\pi: \mathbb{C}^2 \to \mathbb{C}^2/C_b$. The group C_b acts also on $\Sigma(1)$ by means of $A(z,w) = (Az,w)$. The corresponding orbit space is $\Sigma(1)/C_b = \Sigma(b)$, the projection being $\Sigma(1) \to \Sigma(b)$, $(z,w) \mapsto ((z_1^b,z_2^b),w)$. The blowing down $\sigma: \Sigma(1) \to \mathbb{C}^2$ is C_b-equivariant, i.e. $\sigma \circ A = A \circ \sigma$. Therefore a holomorphic mapping $\bar{\sigma}: \Sigma(b) \to \mathbb{C}^2/C_b$ of the orbit spaces is induced. Explicitely it is given by $\bar{\sigma}((z_1,z_2),[w_1:w_2]) = \pi(\zeta_1,\zeta_2)$ where $\zeta_i^b=z_i$ and $\zeta_1 w_2=\zeta_2 w_1$. This $\bar{\sigma}$ is a modification of $\mathbb{C}^2/C_b$ at $\pi(0)$ with the projective line $E = \{(0,w) \in \Sigma_b\}$ as exceptional set. Thus $\bar{\sigma}$ resolves the singularity $\pi(0)$ of $\mathbb{C}^2/C_b$.

Let us consider the other projection $p: \Sigma(b) \to \mathbb{P}_1$ given by $p(z,w)=w$. Its fibres $p^{-1}(w)$ for fixed $w \in \mathbb{P}_1$ are complex lines through 0 in $\mathbb{C}^2$. The line bundle structure of $\Sigma(b)$ is explained below by local charts.

Here is a geometric description of the real analog of $\Sigma(1)$ and the projections σ and p: Let D be a disk in $\mathbb{R}^2$ with centre 0. The real projective line $\mathbb{P}_1$ is a circle line, and $D \times \mathbb{P}_1$ is a solid torus. It contains part of the real $\Sigma(1)$ as a Möbius strip. In order to describe σ the disk D is identified with a fixed transverse slice of the solid torus. Then $\sigma(x)$ is obtained from $x \in \Sigma(1)$ by following the meridian through x within the solid torus until it hits D at $\sigma(x)$. The other projection p goes from $x \in \Sigma(1)$ radially (hence within $\Sigma(1)$) onto $p(x) \in \mathbb{P}_1$ where

$\mathbb{P}_1$ is considered as central core of the solid torus.

The bundle $\Sigma(b)$ is a union of two copies of $\mathbb{C}^2$ (not disjoint). This is seen as follows: Let $\Sigma_i(b) = \{(z,w) \in \Sigma(b): w_i \neq 0\}$. Then $\Sigma(b) = \Sigma_1(b) \cup \Sigma_2(b)$ (union of two open subsets). There are holomorphic charts

$\Sigma_1(b) \approx \mathbb{C}^2$, $(z,w) \mapsto (z_1, w_2/w_1)$, and $\Sigma_2(b) \approx \mathbb{C}^2$, $(z,w) \mapsto (z_2, w_1/w_2)$.

We will call z_1 and z_2 *linear fibre coordinates* and w_2/w_1 resp. w_1/w_2 *affine base coordinates*. The transition from the first chart to the second one is given by

$$(4) \qquad \mathbb{C}\times\mathbb{C}^{\times} \to \mathbb{C}\times\mathbb{C}^{\times} , \quad (u,v) \mapsto (v^b u, v^{-1}) .$$

A mapping $s: \mathbb{P}_1 \to \Sigma(b)$ is called a section if $p\cdot s = id$. There is of course the zero section $s(w)=(0,w)$ which maps $\mathbb{P}_1$ biholomorphically onto E. If $b \geq 1$ there are no other holomorphic sections, because every holomorphic section has the form $s(w)=(s'(w),w)$ where $s': \mathbb{P}_1 \to \mathbb{C}^2$ is a holomorphic mapping. Since $\mathbb{P}_1$ is compact and since the maximum principal holds true for the components of s', it must be constant $s'(w)=(c_1,c_2)$. Therefore $c_1 w_2^b = c_2 w_1^b$ for every $w \in \mathbb{P}_1$ with fixed (c_1,c_2). This implies $(c_1,c_2)=(0,0)$. There are many *local* holomorphic sections which are defined on proper open subsets of $\mathbb{P}_1$, e.g. using the coordinates (u,v) on Σ_1 every holomorphic function $v=s(u)$ "is" a section defined on $\mathbb{P}_1 \setminus \{[0:1]\}$.

Remark: In Algebraic Geometry the bundle $\Sigma(b)$ is usually denoted by $\mathcal{O}(-b)$ because the Euler class is $-b$, see §11. In particular $\Sigma(1)=\mathcal{O}(-1)$ is called the canonical line bundle and $\mathcal{O}(-b)$ is the tensor product $\mathcal{O}(-1)\otimes\ldots\otimes\mathcal{O}(-1)$ with b factors. There are also line bundles $\mathcal{O}(n)=\Sigma(-n)$ for positive n namely $\mathcal{O}(n)$ = dual of $\mathcal{O}(-n)$. The bundle $\mathcal{O}(n)$ is defined by glueing two copies of $\mathbb{C}^2$ according to the transition function $\mathbb{C}\times\mathbb{C}^{\times} \to \mathbb{C}\times\mathbb{C}^{\times}$, $(u,v) \to (v^{-n}u, v^{-1})$. For positive n the bundle $\mathcal{O}(n)$ has non-trivial holomorphic sections. Every line bundle over $\mathbb{P}$ is isomorphic to $\mathcal{O}(n)$ for some $n \in \mathbb{Z}$. But we do not need the $\mathcal{O}(n)$ for $n>0$.

§ 5 Cyclic Quotient Singularities

Different finite cyclic subgroups of $GL(2,\mathbb{C})$ may yield isomorphic orbit varieties. A more precise statement of this fact together with repeated blowings up will yield the resolution of any quotient singularity of a finite cyclic group. Observe that we do not restrict to subgroups of $SL(2,\mathbb{C})$.

Let $A \in GL(2,\mathbb{C})$ generate the finite cyclic subgroup $G < GL(2,\mathbb{C})$. Up to a linear change of coordinates we may assume that

(1) $A = \begin{pmatrix} \zeta & 0 \\ 0 & \eta \end{pmatrix}$, ζ primitive r-th root of unity, η primitive s-th root of unity .

The order of A $(= \# G)$ is the least common multiple $n := \operatorname{lcm}(r,s)$. Let $r \cdot p = n$. The matrix

$$B = \begin{pmatrix} \zeta & 0 \\ 0 & \eta^p \end{pmatrix}$$

has order r . Let $H < GL(2,\mathbb{C})$ denote the group generated by B .

Proposition: *The mapping* $f\colon \mathbb{C}^2 \to \mathbb{C}^2$ *given by* $f(z,w)=(z,w^p)$ *induces a* biholomorphic *mapping* $\bar{f}$ *of the quotient spaces according to the commutative diagram*

$$\begin{array}{ccc} \mathbb{C}^2 & \xrightarrow{\ f\ } & \mathbb{C}^2 \\ \pi \downarrow & & \downarrow \pi \\ \mathbb{C}^2/G & \xrightarrow[\cong]{\ \bar{f}\ } & \mathbb{C}^2/H \end{array} .$$

Proof: We show that $\bar{f}$ is injective. Everything else is straight-forward. Let (z,w) and (z',w') denote points in $\mathbb{C}^2$. If $\pi f(z,w) = \pi f(z',w')$ in $\mathbb{C}^2/H$ there is a q such $z'=\zeta^q z$ and $w'^p=\eta^{pq}w^p$. The second equality implies $w'=\vartheta\eta^q w$ for some p-th root of unity ϑ . Since η^r is a primitive p-th root of unity $\vartheta=\eta^{rl}$ for some l . Let $t:=rl+q$. Then $z'=\zeta^t z$ and $w'=\eta^t w$ i.e. $\pi(z,w)=\pi(z',w')$ in $\mathbb{C}^2/G$.

In the applications of the proposition the rôle of ζ and η will sometimes be interchanged.

Let $C_{n,q} < GL(2,\mathbb{C})$ denote the cyclic subgroup of order n generated by

$$(2) \quad \begin{pmatrix} \zeta & 0 \\ 0 & \zeta^q \end{pmatrix} \quad \text{with } \zeta=e^{2\pi i/n}, \ 1\leq q\leq n, \ \gcd(n,q)=1 .$$

Observe that $C_{n,q} < SL(2,\mathbb{C})$ if and only if $q=n-1$. Let $X_{n,q}$ denote the orbit variety $\mathbb{C}^2/C_{n,q}$. If $n>1$ it is singular at $\pi(0)$ and smooth everywhere else. For $n=1$ we have $X_{1,1}=\mathbb{C}^2$.

Corollary: *The orbit variety of any cyclic subgroup $G < GL(2,\mathbb{C})$ is biholomorphic to some $X_{n,q}$.*

Proof: We may assume that G is generated by A as in (1). If $r=s=n$, a certain power of A has the form (2). It is also a generator of G, hence $\mathbb{C}^2/G = X_{n,q}$ in this case. If $r \neq s$, then $r<n$ or $s<n$ say $r<n$. Then by the proposition $\mathbb{C}^2/G \cong \mathbb{C}^2/H$ where $\#H = r < n = \#G$. This procedure of lowering the order of the group must stop after finitely many steps. When it stops, $\mathbb{C}^2/H = X_{r,q}$ for some divisor r of n and some q between 1 and r with $\gcd(r,q)=1$.

A complex space X is said to have a *cyclic quotient singularity of type* (n,q) at the point x if a neighbourhood of x in X is biholomorphic to a neighbourhood of the singular point in $X_{n,q}$.

We shall construct a modification $X \to X_{n,q}$ which simplifies the singularity, i.e. X will have a cyclic quotient singularity of type (q,r) for some $1\leq r\leq q$ with $\gcd(q,r)=1$: By definition $X_{n,q}$ is the orbit variety of the group generated by $\begin{pmatrix} \zeta & 0 \\ 0 & \zeta^q \end{pmatrix}$, $\zeta=2\pi i/n$, $\gcd(n,q)=1$. But using the proposition $X_{n,q}$ can also be considered as the orbit

variety $\mathbb{C}^2/G$ where G is generated by $A = \begin{pmatrix} \zeta & 0 \\ 0 & \eta\zeta \end{pmatrix}$ with $\eta = e^{2\pi i/q}$, hence $\# G = nq$. Now we proceed as in §4 : The group G acts on $\Sigma = \Sigma(1) = \{(z,w) \in \mathbb{C}^2 \times \mathbb{P}_1 : z_1w_2 = z_2w_1\}$ by means of $A\cdot(z,w) = ((\zeta z_1, \zeta\eta z_2), [w_1 : \eta w_2])$. The blowing down $\sigma: \Sigma \to \mathbb{C}^2$ is G-equivariant, and thus induces a modification $\tilde{\sigma}: \Sigma/G \to \mathbb{C}^2/G \approx X_{n,q}$ at $\pi(0)$ with a projective line as exceptional set. We look for the singularities of Σ/G: Let $\Sigma_i = \{(z,w) \in \Sigma : w_i \neq 0\}$, thus $\Sigma = \Sigma_1 \cup \Sigma_2$. Each Σ_i is G-invariant and biholomophic to $\mathbb{C}^2$. On Σ_1 we fix $w_1 = 1$ and take (z_1, w_2) as coordinates, on Σ_2 we fix $w_2 = 1$ and take (z_2, w_1) as coordinates. With respect to these coordinates the generator A of G acts by $(z_1, w_2) \mapsto (\zeta z_1, \eta w_2)$ on Σ_1 and by $(z_2, w_1) \mapsto (\zeta\eta z_2, \eta^{-1} w_1)$ on Σ_2. Let H_1 be the subgroup generated by $\begin{pmatrix} \zeta & 0 \\ 0 & \eta \end{pmatrix}$ and H_2 the subgroup generated by $\begin{pmatrix} \zeta\eta & 0 \\ 0 & \eta^{-1} \end{pmatrix}$. Then G acts on Σ_i as the linear group H_i on $\mathbb{C}^2$. Using the proposition twice for H_1 we obtain $\mathbb{C}^2/H_1 \cong \mathbb{C}^2$ i.e. Σ_1/G is smooth. Using the proposition for H_2 we obtain $\mathbb{C}^2/H_2 = \mathbb{C}^2/H_3$ with H_3 generated by $\begin{pmatrix} \eta^n & 0 \\ 0 & \eta^{-1} \end{pmatrix}$. By interchanging the coordinates and replacing the generator by its inverse H_3 becomes the group generated by $\begin{pmatrix} \eta & 0 \\ 0 & \eta^r \end{pmatrix}$ with $n = bq - r$ and $0 \leq r \leq q$.
If $r = 0$, then $q = 1$ because $\gcd(n,q) = 1$. In this case $H_3 = \{id\}$ so that $\Sigma_2/G \cong \mathbb{C}^2/H_3 \cong \mathbb{C}^2$ is everywhere smooth, too. If $r > 1$, then $\gcd(q,r) = 1$ hence $\Sigma_2/G \cong \mathbb{C}^2/H_3 = X_{q,r}$. We summarize:

There is a modification $X \to X_{n,q}$ *of the cyclic quotient variety* $X_{n,q} = \mathbb{C}^2/C_{n,q}$ *at the point* $\pi(0)$ *with a projective line as exceptional set so that* X *has at most one singularity. It lies in the exceptional set. If* $q=1$, *then* X *is everywhere smooth. If* $q>1$, *a neighbourhood of the singular point in* X *is biholomorphic to* $X_{q,r}$. *Here* r *is obtained by division with negative remainder:* $n=bq-r$ *for natural numbers* b *and* r *with* $1\leq r\leq q$ *and* $\gcd(q,r)=1$.

§6 The Resolution of Cyclic Quotient Singularities

We have just modified the cyclic quotient singularity $X_{n,q}$ to a complex space X which contains $X_{q,r}$ as open subset and is smooth everywhere else. We repeat this modification at $X_{q,r}$, and insert it into X and so on. After finitely many steps the singularity will disappear, and the resolution of $X_{n,q}$ is obtained. We shall consider the sequence of these modifications, and its final smooth result in more detail.

We begin with a repeated division with negative remainder (a modified Euclidean Algorithm):

$$\frac{n}{q} = b_1 - \frac{q_1}{q} \quad , \quad 0 \leq q_1 < q$$

$$\frac{q}{q_1} = b_2 - \frac{q_2}{q_1} \quad , \quad 0 \leq q_2 < q_1 \quad .$$

After finitely many steps $q_s=1$, and the final equality is

$$\frac{q_{s-1}}{q_s} = b_{s+1} \quad .$$

Since $\gcd(n,q)=1$, we have $\gcd(q_i,q_{i+1})=1$ at every step $(q_o=q)$. There is a sequence of modifications

(1) $X_{n,q} \leftarrow Y_{q,q_1} \leftarrow Y_{q_1,q_2} \leftarrow \dots \leftarrow Y_{q_{s-1},1} \leftarrow Y$.

The exceptional set of each modification is a projective line. Each $Y_{r,s}$ is smooth up to one singularity. It lies in the exceptional set of the preceding modification, and is the point at which $Y_{r,s}$ is modified in the following step. The singularity of $Y_{r,s}$ has a neighbourhood which is biholomorphic to the cyclic quotient singularity $X_{r,s}$. The final complex space Y is everywhere smooth. The composition $\sigma: Y \to X_{n,q}$ is a resolution of the singularity of $X_{n,q}$.

<u>Theorem</u>: *The smooth complex surface* Y *which resolves the cyclic quotient singularity* $X_{n,q}$ *is a union of* $s+2$ *copies of* $\mathbb{C}^2$,

$$Y = \mathbb{C}^2_{(1)} \cup \ldots \cup \mathbb{C}^2_{(s+2)} ,$$

in such a way that

$$\mathbb{C}^2_{(i)} \cup \mathbb{C}^2_{(i+1)} = \Sigma(b_i)$$

is the line bundle. Hence also

$$Y = \Sigma(b_1) \cup \ldots \cup \Sigma(b_{s+1})$$

with

$$\Sigma(b_{i-1}) \cap \Sigma(b_i) = \mathbb{C}^2_{(i)} .$$

There are coordinates (x_i, y_i) *of* $\mathbb{C}^2_{(i)}$ *such that* x_i *is an affine base coordinate of* $\Sigma(b_{i-1})$ *and a linear fibre coordinate of* $\Sigma(b_i)$ *and* y_i *is a linear fibre coordinate of* $\Sigma(b_{i-1})$ *and an affine base coordinate of* $\Sigma(b_i)$.

The exceptional set E *of the resolution* $\sigma: Y \to X_{n,q}$ *is the union of the zero sets* E_i *of* $\Sigma(b_i)$,

$$E = E_1 \cup \ldots \cup E_{s+1} ,$$

hence a union of $s+1$ *projective lines. Here* E_{i-1} *and* E_i *intersect transversely in one point and* $E_i \cap E_j = \emptyset$ *for* $i-j \neq \pm 1$

Proof: Let $\pi: \mathbb{C}^2 \to X_{n,q}$ denote the canonical projection. The points of $X_{n,q}$ will be denoted by $[z,w] = \pi(z,w)$. This notation will also be used with other indices $X_{q,r}$ etc. We look at the first two modifications of the sequence (1), but use a different notation avoiding double indices:

$$X_{n,q} \overset{\sigma}{\leftarrow} Y_{q,r} \overset{\tau}{\leftarrow} Y_{r,s} \quad \text{with } n=bq-r \qquad (b=b_1) .$$

We redo the construction at the end of the preceding §5 in more detail:

$$Y_{q,r} = \mathbb{C}^2_{(1)} \cup X_{q,r}$$

with the following identifications

(2) $\mathbb{C}^2_{(1)} \ni (x_1,y_1) \sim [z,w] \in X_{q,r}$ if and only if
$z^n w = x_1$ and $z^q y_1 = 1$.

The blowing down σ is given by

(3a) $\sigma(x_1,y_1) = [\xi_1, \xi_1^q y_1]$ where $\xi_1^n = x_1$,

(3b) $\sigma([z,w]) = [\eta z, \eta^q]$ where $\eta^n = w$.

The exceptional set E of σ consists of

$$E \cap \mathbb{C}^2_{(1)} = \{(0,y_1)\} \text{ and } E \cap X_{q,r} = \{[z,0]\} .$$

In the next step $X_{q,r}$ is blown up in the same way, $\tau\colon \mathbb{C}^2_{(2)} \cup X_{r,s} \to X_{q,r}$. We need only know the image of (x_2,y_2):

$$\tau(x_2,y_2) = [\xi_2, \xi_2^r y_2] \quad \text{where} \quad \xi_2^q = x_2 ,$$

compare (3a). We extend to the modification

$$\tau\colon Y_{r,s} := \mathbb{C}^2_{(1)} \cup \mathbb{C}^2_{(2)} \cup X_{r,s} \to \mathbb{C}^2_{(1)} \cup X_{q,r} = Y_{q,r}$$

by $\tau|\mathbb{C}^2_{(1)} = \mathrm{id}$. Thus $(x_1,y_1) \in \mathbb{C}^2_{(1)}$ and $(x_2,y_2) \in \mathbb{C}^2_{(2)}$ are identified in $Y_{r,s}$ if and only if $(x_1,y_1) \sim \tau(x_2,y_2)$ by means of (2), i.e. if and only if

(4) $x_2 y_1 = 1$ and $x_2^b y_2 = x_1$.

The comparism with the transition between the two charts of the line bundle $\Sigma(b)$ shows that $\mathbb{C}^2_{(1)} \cup \mathbb{C}^2_{(2)} = \Sigma(b)$. In order to see the intersection of two consecutive line bundles in Y we go one step further:

$$\mathbb{C}^2_{(1)} \cup \mathbb{C}^2_{(2)} \cup \mathbb{C}^2_{(3)} = \Sigma(b_1) \cup \Sigma(b_2) ,$$

more precisely

$$\mathbb{C}^2_{(1)} \cup \mathbb{C}^2_{(2)} = \Sigma(b_1) , \ \mathbb{C}^2_{(2)} \cup \mathbb{C}^2_{(3)} = \Sigma(b_2) , \ \Sigma(b_1) \cap \Sigma(b_2) = \mathbb{C}^2_{(2)} .$$

The claim about the coordinates (x_i,y_i) , here $i=2$, follows from the transition (4) and the corresponding transition between $\mathbb{C}^2_{(2)}$ and $\mathbb{C}^2_{(3)}$ which is given by

$$x_3 y_2 = 1 \text{ and } x_3^{b_2} y_3 = x_2 .$$

It suffices to consider the intersection of consecutive line bundles in Y because $\Sigma(b) \cap X_{r,s} \subset \mathbb{C}^2_{(2)}$ and because further line bundles $\Sigma(b_i)$, $i>1$, are contained in the modification of $X_{r,s}$.

The exceptional set of the composed modification

$$\sigma\tau : Y_{r,s} \to X_{n,q}$$

consists of the zero section of $\Sigma(b)$ and of

$$\{[z,0]\} \subset X_{r,s} \quad .$$

This yields the claim about the exceptional set E of the total modification $Y \to X_{n,q}$.

§ 7 The Cotangent Action

Let $G < SL(2,\mathbb{C})$ be a finite subgroup. We consider the line bundle $\Sigma(1) = \{(z,w) \in \mathbb{C}^2 \times \mathbb{P}_1;\ z_1 w_2 = z_2 w_1\}$ and its zero section $E = \{(0,w) \in \Sigma(1)\} \cong \mathbb{P}_1$. The group G acts on $\Sigma(1)$ by means of $g\cdot(z,w)=(gz,gw)$ for $g \in G$. The zero section E is G-invariant. The blowing down $\sigma\colon \Sigma(1) \to \mathbb{C}^2$, $\sigma(z,w)=z$, is G-equivariant. The induced mapping $\bar{\sigma}\colon \Sigma(1)/G \to \mathbb{C}^2/G$ is a modification with exceptional set $E/G \approx \mathbb{P}_1$. This isomorphism follows from

<u>Proposition 1</u>: *Let the finite group* G *act (linearly) on* $\mathbb{P}_1$. *Then the orbit space* $\mathbb{P}_1/G$ *is biholomorphic to* $\mathbb{P}_1$.

Proof: Let V denote the 2-dimensional vector space of the principal forms, see II § 8. Let $\mathbb{P}(V)$ denote the corresponding projective line. For every $x \in \mathbb{P}_1$ let f_x denote a principal form of the principal divisior through x. Then f_x is uniquely determined up to a constant factor. Hence $\mathbb{P}_1 \to \mathbb{P}(V)$, $x \mapsto f_x$, is a well defined surjective holomorphic mapping. The points x and y of $\mathbb{P}_1$ belong to the same orbit if and only if $f_x = f_y$. Therefore $\mathbb{P}_1 \to \mathbb{P}(V)$ yields a biholomorphic mapping $\mathbb{P}_1/G \approx \mathbb{P}(V)$.

<u>Remark</u>: The assumption that G acts linearly is superfluous, see p.34.

From now on G is assumed to be binary, i.e. the kernel $C_2=\{\pm E\}$ of $\rho\colon SL(2,\mathbb{C}) \to PGL(2,\mathbb{C})$ is a subgroup of

G , see II§1. The orbit projection of the C_2-action on $\Sigma(1)$ is $f\colon \Sigma(1) \to \Sigma(2)$ given by $f((z_1,z_2),w)=((z_1^2,z_2^2),w)$, see §4. The action of G on $\Sigma(1)$ induces a unique action of $\Gamma=\rho(G)$ on $\Sigma(2)$ such that $f\circ g = \rho(g)\circ f$ for $g\in G$. The projection f induces a *biholomorphic* mapping $\bar f$ of the corresponding orbit spaces, see the following diagram:

$$\begin{array}{ccccc}
\Sigma(2) & \xleftarrow{f} & \Sigma(1) & \xrightarrow{\sigma} & \mathbb{C}^2 \\
\pi\downarrow & & \pi\downarrow & & \pi\downarrow \\
\Sigma(2)/\Gamma & \underset{\approx}{\xleftarrow{\bar f}} & \Sigma(1)/G & \xrightarrow{\bar\sigma} & \mathbb{C}^2/G
\end{array}$$

For later reference we summarize the result obtained so far:

Proposition 2: *There is a modification* $\Sigma(2)/\Gamma \to \mathbb{C}^2/G$ *with exceptional set* $E/\Gamma \approx \mathbb{P}_1$ *where* E *is the zero section of* $\Sigma(2)$.

The following proposition brings the cotangent bundle into play:

Proposition 3: *The line bundle* $\Sigma(2)$ *is canonically isomorphic to the cotangent bundle* $T^*\mathbb{P}_1$.

Proof: The local holomorphic sections of $T^*\mathbb{P}_1$ are the holomorphic forms $h(w)d(w_1/w_2) = -h(w)\left(\frac{w_1}{w_2}\right)^2 d(w_2/w_1)$ with h a holomorphic function. It suffices to establish a canonical isomorphism between the local holomorphic sections of $\Sigma(2)$ and $T^*\mathbb{P}_1$: Let the section of $\Sigma(2)=\{(z,w)\in\mathbb{C}^2\times\mathbb{P}_1 : z_1w_2^2=z_2w_1^2\}$ be given by $w \mapsto (z_1(w),z_2(w))$ with holomorphic functions z_1 and z_2 . The corresponding form is $z_2(w)d(w_1/w_2) = -z_1(w)d(w_2/w_1)$.

The isomorphism $\Sigma(2) \cong T^*\mathbb{P}_1$ transfers the action of Γ from $\Sigma(2)$ to $T^*\mathbb{P}_1$. This action is a special case of a more general construction which we now describe: Let $f\colon X \to Y$ be a holomorphic mapping between two smooth complex curves. Every (locally defined) holomorphic form ω on Y can be lifted to the form $f^*\omega$ on X: If y is a holomorphic coordinate on Y we write $\omega = h\,dy$ with a holomorphic function h and obtain $f^*\omega = (h\circ f)d(y\circ f)$. We associate a *cotangent mapping* $f_*: T^*X \to T^*Y$ with a locally biholomorphic mapping $f: X \to Y$ as follows: Let $c \in T^*_x X$ be given. Choose a holomorphic form ω defined in a neighbourhood U of x such that $\omega(x) = c$. Let U be so small that f maps U biholomorphically onto the open subset $V \subset Y$. Let $g: V \to U$ be the inverse mapping of $f|U$. It induces the form $g^*\omega$ on V. Then $f_*(c) = (g^*\omega)(f(x))$. This f_* is a well defined holomorphic mapping such that the following diagram is commutative

$$\begin{array}{ccc} T^*X & \xrightarrow{f_*} & T^*Y \\ p\downarrow & & \downarrow p \\ X & \xrightarrow{f} & Y \end{array}$$

We have $(h\circ f)_* = h_*\circ f_*$ for the composition of two locally biholomorphic mappings f and h. Here are two examples of cotangent mappings:

1st Example: Let $f: \mathbb{C} \to \mathbb{C}$ be given by $f(z) = a\cdot z$, $a \in \mathbb{C}^\times$. Then $T^*\mathbb{C} = \mathbb{C}\times\mathbb{C}$ (the base is the first factor, the fibre the second one) and f_* is given by the matrix

$$\begin{pmatrix} a & 0 \\ 0 & a^{-1} \end{pmatrix}$$

2nd Example: Let $f: \mathbb{C}^\times \to \mathbb{C}^\times$ be given by $f(z) = z^n$. Then $T^*\mathbb{C}^\times = \mathbb{C}^\times \times \mathbb{C}$ and f_* is given by $f_*(z,w) = (z^n, \frac{1}{n}z^{1-n}w)$.

The action of the element $\gamma \in \Gamma$ on $\mathbb{P}_1$ yields the associate cotangent action γ_* on $T^*\mathbb{P}_1$. Under the canonical isomorphism $T^*\mathbb{P}_1 \cong \Sigma(2)$ the associate action of Γ on $T^*\mathbb{P}_1$ corresponds to the action of Γ on $\Sigma(2)$ which is induced by the action of G on $\Sigma(1)$.

The results of this section are summarized in

<u>Proposition 4</u>: *Let* $\Gamma < PGL(2,\mathbb{C})$ *be a finite subgroup, let* $G < SL(2,\mathbb{C})$ *be its inverse image. There is the cotangent action of* Γ *on* $T^*\mathbb{P}_1$ *associated to the action of* Γ *on* $\mathbb{P}_1$. *Its orbit space is the total space of a modification* $T^*\mathbb{P}_1/\Gamma \to \mathbb{C}^2/G$ *with exceptional set* $E/\Gamma \approx \mathbb{P}_1$, *where* $E \subset T^*\mathbb{P}_1$ *denotes the zero section.*

§8 Line Bundles with Singularities

In this section $\Gamma < PGL(2,\mathbb{C})$ *is not cyclic.* The orbit space $T^*\mathbb{P}_1/\Gamma$ sits in the commutative diagram

$$\begin{array}{ccc} T^*\mathbb{P}_1 & \xrightarrow{\pi_*} & T^*\mathbb{P}_1/\Gamma \\ p\downarrow & & \downarrow\bar{p} \\ \mathbb{P}_1 & \xrightarrow{\pi} & \mathbb{P}_1 \approx \mathbb{P}_1/\Gamma \end{array}$$

where π and π_* denote the orbit projection, p is the bundle projection and $\bar{p}$ is the induced mapping. The projective line $\mathbb{P}_1$ is identified with the two-sphere S^2 and Γ is considered as subgroup of $SO(3)$, see II § 1. The surface of the regular solid corresponding to Γ is projected radially onto $S^2 \approx \mathbb{P}_1$. There are three exceptional orbits: Σ_1 = {mid-edge points}, Σ_2 = {face centers}, and Σ_3 = {vertices}, see I§6. Considered as points of the orbit space $\mathbb{P}_1 = \mathbb{P}_1/\Gamma$ they are denoted by y_1, y_2, and y_3 respectively, hence $\Sigma_i = \pi^{-1}(y_i)$. Off the exceptional orbits $\pi: \mathbb{P}_1 \setminus (\Sigma_1 \cup \Sigma_2 \cup \Sigma_3) \to \mathbb{P}_1 \setminus \{y_1, y_2, y_3\}$ is locally biholomorphic. Thus a cotangent mapping $\pi_*: T^*(\mathbb{P}_1 \setminus (\Sigma_1 \cup \Sigma_2 \cup \Sigma_3)) \to T^*(\mathbb{P}_1 \setminus \{y_1, y_2, y_3\})$ is associated. It is the orbit projection of the Γ-action on $T^*\mathbb{P}_1 \setminus p^{-1}(\Sigma_1 \cup \Sigma_2 \cup \Sigma_3) = T^*(\mathbb{P}_1 \setminus (\Sigma_1 \cup \Sigma_2 \cup \Sigma_3))$.

This implies

(1) $(T^*\mathbb{P}_1/\Gamma)\setminus\bar{p}^{-1}\{y_1,y_2,y_3\}=T^*(\mathbb{P}_1\setminus\{y_1,y_2,y_3\})=T^*\mathbb{P}_1\setminus p^{-1}\{y_1,y_2,y_3\}$.

We investigate a neighbourhood of the fibre $\bar{p}^{-1}(y_i)\subset T^*\mathbb{P}_1/\Gamma$: Let $c\in\Sigma_i\subset\mathbb{P}_1$ be a mid-edge point (i=1) or a face center (i=2) or a vertex (i=3) . The isotropy subgroup Γ_c is cyclic of order $n=n_i$ where n_i is given in the table of I§6 . A generator of Γ_c is a $2\pi/n$-rotation. There is a local holomorphic coordinate z defined in a neighbourhood U of c in $\mathbb{P}_1$ such that $z(c)=0$, $z(U)\subset D$ = unit disk, and the generator of Γ_c acts by $z\mapsto e^{2\pi i/n}\cdot z$, compare §3, proposition 2. Let w be a linear fibre coordinate defined on $p^{-1}(U)\subset T^*\mathbb{P}_1$. According to the first example of a cotangent mapping (see §7) the generator of Γ_c acts by $(z,w)\mapsto(e^{2\pi i/n}\cdot z,e^{-2\pi i/n}\cdot w)$ on $p^{-1}(U)$. Now $\pi_*(p^{-1}(U)) = p^{-1}(U)/\Gamma_c$ is a neighbourhood of $\bar{p}^{-1}(y_i)$ in $T^*\mathbb{P}_1/\Gamma$. Therefore $T^*\mathbb{P}_1/\Gamma$ has a cyclic quotient singularity of type (n_i,n_i-1) at $\bar{s}(y_i)$. Here $\bar{s}\colon\mathbb{P}_1/\Gamma\to T^*\mathbb{P}_1/\Gamma$ is induced from the zero section $s\colon\mathbb{P}_1\to T^*\mathbb{P}_1$.

The orbit space $T^*\mathbb{P}_1/\Gamma$ will be obtained from the cotangent bundle $T^*\mathbb{P}_1$ be removing three fibres $p^{-1}\{y_1,y_2,y_3\}$ and inserting instead three neighbourhoods $p^{-1}(U)/\Gamma_c = (U\times\mathbb{C})/\Gamma_c$ which have been described above. This can be done more generally and yields a "line bundle with singularities":
Let $p\colon L\to S$ denote a line bundle (in the usual sense) over the smooth complex surface S . Let $b\in S$. We construct a complex space L' by *inserting a cyclic quotient singularity* of type (n,q) into the fibre over b: Let t be a holomorphic coordinate defined in a neighbourhood U of b in S such that the unit disk $D=\{z\in\mathbb{C}\colon |z|<1\}$ is contained in the image of t and such that $t(b)=0$. Let u be a linear fibre coordinate defined on $p^{-1}(U)$. Then (t,u) is a holomorphic chart in $p^{-1}(U)$. The cyclic subgroup of $GL(2,\mathbb{C})$ generated by

$$\begin{pmatrix} e^{2\pi i/n} & 0 \\ 0 & e^{2\pi iq/n} \end{pmatrix}$$

acts on $D\times\mathbb{C}$. Let $\pi\colon D\times\mathbb{C} \to X'_{n,q}$ be the corresponding orbit projection. We observe that $X'_{n,q}$ is a neighbourhood of the singularity in $X_{n,q}=\mathbb{C}^2/C_{n,q}$, see §5 (2). We define the *line bundle with singularity*

$$L' = (L\setminus p^{-1}(b)) \cup X'_{n,q}$$

by the identification

(1) $L\setminus p^{-1}(b) \ni x \sim \pi(z,w) \in X'_{n,q} \Leftrightarrow t(x)=z^n$ and $u(x)=z^{-q}w$.

The given projection $p\colon L\setminus p^{-1}(b) \to S\setminus\{b\}$ extends to $p\colon L' \to S$ by $p\colon X'_{n,q} \to U$, $t(p\pi(z,w))=z^n$. Then $L\setminus p^{-1}(b) = L'\setminus p^{-1}(b)$. The given zero section $s\colon S\setminus\{b\} \to L\setminus p^{-1}(b)$ extends to $s\colon S \to L'$ by $s\colon U \to X'_{n,q}$, $s(x)=\pi(z,0)$ where $z^n=t(x)$. This s maps S biholomorphically onto its image $s(S)$. This image is also called the zero section of L' . There is one singularity in L' . It lies in the fibre of b and is a cyclic quotient singularity of type (n,q) .

We may repeat this construction, and insert cyclic quotient singularities into several different fibres.

<u>An Example</u>: In the resolution of the cyclic quotient singularity $X_{n,q}$, see §6 , the first step has been the modification $X_{n,q} \leftarrow Y_{q,r}$. Here $Y_{q,r}$ is obtained from the line bundle $\Sigma(b)$ be inserting a cyclic quotient singularity of type (q,r) into one fibre.

We apply the insertion of singular fibres to $T^*\mathbb{P}_1$: Let t be a holomorphic coordinate defined in a neighbourhood of y_i in $\mathbb{P}_1$ such that near c and $\pi(c)=y_i$ the orbit projection $\pi\colon \mathbb{P}_1 \to \mathbb{P}_1$ is given by $t=z^n$. According to the second example of a cotangent mapping, see §7 , we have $\pi_*(z,w) = (z^n,\frac{1}{n}z^{1-n}w)$ for $z\neq 0$. The factor $1/n$ means an unimportant linear change of the fibre coordinate w . Thus near the fibre over y_i the quotient space $T^*\mathbb{P}_1/\Gamma$ is obtained from $T^*\mathbb{P}_1$ by inserting a quotient singularity of type $(n,n-1)$ into the fibre over y_i. Since $T^*\mathbb{P}_1/\Gamma$ and $T^*\mathbb{P}_1$ coincide off the fibres over y_1, y_2 and y_3, see (1), we obtain:

Proposition: *The orbit space* $T^*\mathbb{P}_1/\Gamma$ *of the cotangent action described in §7 Proposition 4 is obtained from* $T^*\mathbb{P}_1$ *by inserting three cyclic quotient singularities of types* (n_i,n_i-1) *with* $i=1,2,3$ *into three different fibres. The values of* n_i *are given by*

type of Γ	n_1	n_2	n_3
q-*dihedral*	2	2	q
tetrahedral	2	3	3
octahedral	2	3	4
icosahedral	2	3	5

§9 The Resolution of Non-Cyclic Quotient Singularities

Let $G < SL(2,\mathbb{C})$ be a finite non-cyclic subgroup. The final results of §7 and §8 yield a *modification* $\tau: Z \to \mathbb{C}^2/G$ *at the singular point such* **that** Z *is obtained from the cotangent bundle* $T^*\mathbb{P}_1 = \Sigma(2)$ *by inserting cyclic quotient singularities of types* (n_1,n_1-1), (n_2,n_2-1), *and* (n_3,n_3-1) *into three different fibres. The exceptional set is the zero section of* Z, *which is a projective line. The numbers* (n_1,n_2,n_3) *are given in the table at the end of* §8.

In order to resolve the singularity $\mathbb{C}^2/G$ completely it will suffice to resolve the cyclic quotient singularities of Z. We shall do this more generally for the bundle L' which has been obtained from the line bundle $p: L \to S$ by inserting a cyclic quotient singularity of type (n,q), see §8. A neighbourhood of this inserted singularity is an open subset $X'_{n,q}$ of $X_{n,q}$. Let $\sigma: Y = \mathbb{C}^2_{(1)} \cup \ldots \cup \mathbb{C}^2_{(s+2)} \to X_{n,q}$ be the resolution which has been constructed in §6 . We restrict to $Y' = \sigma^{-1}(X'_{n,q})$ and particularly to $U_{(1)} = Y' \cap \mathbb{C}^2_{(1)} = \{(x_1,y_1) \in \mathbb{C}^2_{(1)}: |x_1| < 1\}$.

<u>Proposition:</u> *A resolution* $\tau: X \to L'$ *of the singularity of* L' *is obtained as follows:*

$$X = L \cup Y' \quad \text{(union of open subsets)}$$

with the identifications
$L \ni x \sim y \in Y' \iff x \in p^{-1}(U),\ y \in U_{(1)}$ *and* $t(x) = x_1(y),\ u(x) = y_1(y)$.

The mapping τ *is given by:*

If $x \in L \setminus p^{-1}(b)$, *then* $\tau(x) = x \in L' \setminus p^{-1}(b)$,
if $y \in Y'$, *then* $\tau(y) = \sigma(y) \in X'_{n,q} \subset L'$.

Observe that the fibre $p^{-1}(b) \subset L$ is also contained in Y' . The reader will check that both definitions of τ coincide on $(L \setminus p^{-1}(b)) \cap Y'$ using (1) and § 6(3a) .

Proof of the proposition: In the beginning of § 4 we described how to insert modifications into complex spaces. Using this general procedure the modification X of L' is obtained from $L' = (L \setminus p^{-1}(b) \cup X'_{n,q}$ by replacing $X'_{n,q}$ by Y' and identifying $L \setminus p^{-1}(b) \ni x \sim y \in Y'$ if and only if x and $\sigma(y) \in X'_{n,q}$ are identified according to §8(1). Using the de - scription of $\sigma | \mathbb{C}^2_{(1)}$ given in §6(3a) we obtain $x \sim y$ $t(x) = x_1(y)$ and $u(x) = y_1(y)$. Therefore we need not remove the fibre $p^{-1}(b)$ in order to identify L and Y' (in contrast to the identification §8(1) between L and $X'_{n,q}$) .

<u>Corollary:</u> *The exceptional set of* $\tau: X \to L'$ *coincides with the exceptional set* $E = E_1 \cup \ldots \cup E_{s+1}$ *of* $\sigma: Y \to X_{n,q}$, *see the theorem in* § 6 . *Here* E_1 *intersects the zero section* S *of* $L \subset X$ *transversely in one point and* $E_i \cap S = \emptyset$ *for* $i \geq 2$.

The proposition and its corollary extend in an obvious way to line bundles with several cyclic quotient singularities inserted into different fibres. The different Y's don't intersect each other.

When we resolve the three singularities of Z in this way a resolution $\varphi: X \to \mathbb{C}^2/G$ is obtained. It looks as follows: Let $\sigma_i: Y_i' \to X'_{n_i,n_i-1}$ be the resolution constructed in §6 and restricted to the open subset $X'_{n_i,n_i-1} \subset X_{n_i,n_i-1}$. Here Y_i' is a neighbourhood of the exceptional set $F_i \subset Y_i$ of σ_i. The total space is

$$X=\Sigma(2) \cup Y_1' \cup Y_2' \cup Y_3' .$$

The identifications between $\Sigma(2)=T^*\mathbb{P}_1$ and each Y_i' are given by the proposition above. We have $Y_i' \cap Y_j' = \emptyset$ for $i \neq j$. The exceptional set of φ is $E=E_o \cup F_1 \cup F_2 \cup F_3$ where E_o is the zero section of $\Sigma(2)$. According to the theorem in §6 $F_i=E_{i,1} \cup \ldots \cup E_{i,s_i+1}$ is a union of projective lines. The zero section E_o intersects $E_{i,1}$ transversely and does not meet $E_{i,j}$ for $j \geq 2$.

§10 Plumbed Surfaces

In order to have a more convenient description of the resolution $X \to \mathbb{C}^2/G$ of a quotient singularity the *plumbing of surfaces* according to a weighted tree will be introduced.

A graph T is called a *tree* if it can be built up in finitely many steps $T_1 \subset T_2 \subset \ldots \subset T_k = T$ such that T_1 consists of one vertex, called 1, and no edge and such that T_{i+1} is obtained from T_i by adding an edge which joins a vertex of T_i to the new vertex $i+1$, see the figure 1.

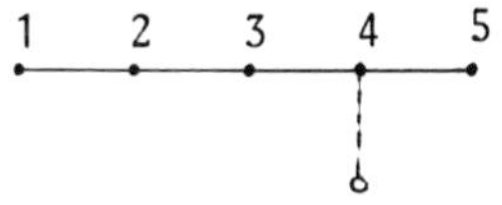

------ new edge

o new vertex

Figure 1

The figure 2 shows all possible trees with at most five vertices

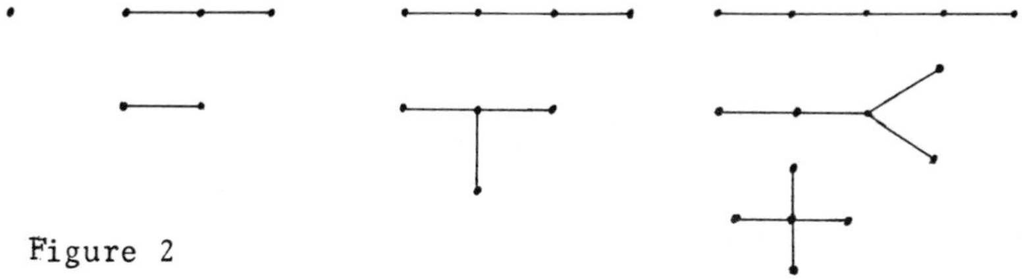

Figure 2

The tree is called weighted, if an integer b_i is assigned to every vertex i .

We assign a weighted tree (T,b) to the resolution $X \to \mathbb{C}^2/G$ as follows: Let $E=E_1\cup\ldots\cup E_k \subset X$ denote the exceptional set. It consists of projective lines E_i . The graph T has one vertex i for each E_i . Two vertices i and j are connected by an edge (briefly i—j) if and only if $E_i \cap E_j \neq \emptyset$. In order to determine the weights b_i we observe that each E_i has a neighbourhood $X_i \subset X$ such that the pair (X_i,E_i) is biholomorphic to $(N(b_i),S)$ where $N(b_i)$ is a neighbourhood of the zero section $S \approx \mathbb{P}_1$ in $\Sigma(b_i)$. The vertex is is weighted by b_i . We shall see in §11 and §12 that b_i is well defined, because $-b_i$ is the self intersection number of E_i in X .

Theorem 1: *For the cyclic quotient singularity of type* (n,q) *the weighted tree of the resolution constructed in* §6 *is obtained from the repeated division with negative remainder:*

This follows immediately from §6 .

<u>Theorem 2</u>: *For finite subgroups* $G < SL(2,\mathbb{C})$ *the resolutions of* $\mathbb{C}^2/G$ *constructed in* §6-§9 *have the following weighted* ***trees:***

type of G	tree	*number of vertices*
cyclic of order n		n-1
binary q-*dihedral*		q+2
binary tetrahedral		6
binary octahedral		7
binary icosahedral		8

All weights are = 2 .

Proof: The repeated division of $(n,n-1)$ has $s+1=n-1$ terms. All b_i's are = 2 . This yields the graphs of the cyclic groups as particular cases of theorem 1. The weighted trees for the non-cyclic subgroups G follow from the description of X and E at the end of §9 . The triple vertex of the tree corresponds to the zero section E_o of $\Sigma(2)=T^*\mathbb{P}_1$ and the three branches correspond to the exceptional sets F_i $(i=1,2,3)$ of the cyclic quotient singularities of type (n_i,n_i-1) . The n_i's are listed at the end of §8 .

A smooth complex surface X is called *plumbed* according to the weighted tree (T,b) with k vertices if the following conditions are fullfilled:

(1) $X=X_1\cup\ldots\cup X_k$ is a union of open subsets X_i , one for each vertex i , such that X_i is biholomorphic to a tubular neighbourhood of the zero section in $\Sigma(b_i)$.

Tubular neighbourhood of the zero section in the line bundle $L \to S$ is defined as follows: A continuous function $\rho: L \to \mathbb{R}$ is called a *norm* if the restriction of ρ to each fibre is a norm in the usual sense.

(Remember that the fibres are 1-dimensional complex vector spaces). Let $\varepsilon > 0$. Then $\{x \in L: \rho(x) < \varepsilon\}$ is a tubular neighbourhood of the zero section.

(2) If the vertices i and j are not connected by an edge, $X_i \cap X_j = \emptyset$.

(3) If i and j are connected by an edge, there is a holomorphic chart (z,w) defined on $X_i \cap X_j$ which maps $X_i \cap X_j$ biholomorphically onto $D \times D$ (D= open unit disk in $\mathbb{C}$) such that
z is a holomorphic base coordinate of X_i and a linear fibre coordinate of X_j
and
w is a linear fibre coordinate of X_i and a holomorphic base coordinate of X_j .

Here base and fibre coordinate refer to $X_i \subset \Sigma(b_i)$ and $X_j \subset \Sigma(b_j)$. The condition (3) means that X_i and X_j are plumbed crosswise, see the figure.

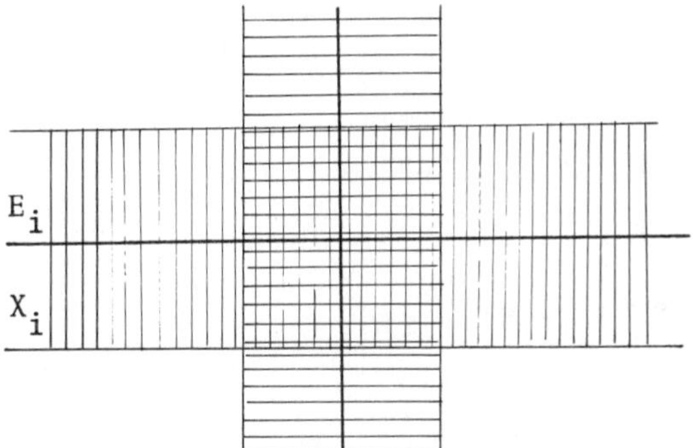

Let $E_i \subset X_i$ denote the zero section. It is a complex line. If $i \not- j$, then $E_i \cap E_j = \emptyset$, if $i - j$, then E_i and E_j intersect transversally in one point. The union $E = \bigcup_{i=1}^{k} E_i$ is called the *core* of X .

Theorem 3: *Let* X *be the resolution of a quotient singularity, let* (T,b) *its weighted tree. A suitable neighbourhood of the exceptional set* E *in* X *is a surface* Y *which is plumbed according to* (T,b) *with core* E . *There are arbitrary small such neighbourhoods.*

Proof: The resolution X has been described in §6 for the cyclic quotient singularities and at the end of §9 for the non-cyclic quotient singularities. The theorem is another version of these descriptions.

§11 Intersection Numbers

Intersection numbers are considered from the viewpoint of differential topology following the presentation in Hirsch's book chapter 5, section 2. All manifolds will be differentiable oriented, without boundary.

Let W be a manifold of dimension $m+n$ and $N \subset W$ a closed submanifold of dimension n . Let M be a manifold of dimension m . If the map $f: M \to W$ is differentiable at the point $x \in M$, the differential $f_*: T_xM \to T_{f(x)}W$ is a linear map between the tangent spaces. The map f is called *transverse to* N *at* $x \in f^{-1}(N)$ if it is differentiable at x and $f_*(T_xM)+T_{f(x)}N = T_{f(x)}W$. In this case the *local intersection number*

$$\#_x(f,N) = \pm 1$$

is defined as follows: Let $(v_1,..,v_m)$ be a positively oriented basis of T_xM , and let $(w_1,..,w_n)$ be a positively oriented basis of $T_{f(x)}N$. Then $\mathcal{B} = (f_*(v_1),..,f_*(v_m), w_1,..,w_n)$ is a basis of $T_{f(x)}W$, and $\#_x(f,N) = 1$ if and only if $\mathcal{B}$ is positively oriented.

The continuous map $f: M \to W$ is called *transverse* to N if f is transverse at every point $x \in f^{-1}(N)$. If M is compact, $f^{-1}(N)$ is finite, and the (total) *intersection number*

$$\#(f,N) = \sum_{x \in f^{-1}(N)} \#_x(f,N) \tag{1}$$

is defined. If $f,g: M \to W$ are homotopic and transverse to N, then $\#(f,N)=\#(g,N)$. Every continuous map $f: M \to W$ is homotopic to a map g which is transverse to N. Therefore the intersection number $\#(f,N)=\#(g,N)$ is well defined for continuous maps f.

If $M\subset W$ is a compact submanifold, the inclusion $j: M \hookrightarrow W$ is used in order to define the intersection number $\#(M,N)=\#(j,N)$ In particular for $m=n$ there is the *self-intersection number* $\#(M,M)$ of M in W.

Let $p: L \to S$ be a complex line bundle over the compact complex curve S. Then L is a complex surface. The *zero section* $s_o: S \to L$ assigns to every $x\in S$ the zero of the fibre $p^{-1}(x)$. Its image $L_o=s_o(S)$ is also called *zero section*. It is a compact curve in L. Remember that L and $S\approx L_o$ are oriented real manifolds of dimensions 4 and 2 respectively, see the end of III§13. Every continuous section $s: S \to L$ is homotopic to s_o, and thus can be used in order to calculate the self-intersection number

$$e(L) = \#(L_o,L_o) = \#(s,L_o) .$$

This is also called the *Euler number* or the *first Chern number* of L. In the proposition below $e(L)$ will be calculated by means of a *meromorphic* section s. "Meromorphic" means: For any local holomorphic base coordinate t and linear fibre coordinate u the section s is given by a meromorphic function $u(t)$. The poles and zeros of $u(t)$ and their orders do not depend on the choice of t and u. A meromorphic section is not continuous at its poles.

Proposition: *If the line bundle* L *admits a meromorphic section with* k *simple zeros and* n *simple poles, the Euler number* $e(L)$ *equals* $k-n$.

Remark 1: The proposition remains true if the zeros and poles are not simple. They must be counted with their orders.

Remark 2: We must admit meromorphic sections because the line bundles $\Sigma(n)$, we are especially interested in, have no global holomorphic sections $\neq s_o$, see §4.

Proof of the proposition: Let $c\in S$ be a simple pole of s. There are a holomorphic base coordinate t defined in a neighbourhood of c and a linear fibre coordinate u such that $t(c)=0$, and such that s is locally given by $u(t)=1/t$. We replace s by a section s^* which is continuous at c, and coincides with s outside a neighbourhood of c: Let $\varepsilon>0$ be small. Define $u^*(t)=\varepsilon^{-2}\bar{t}$ for $|t|\leq\varepsilon$ and $u^*(t)=1/t$ for $|t|\geq\varepsilon$ where $\bar{t}$ denotes the complex conjugate of t. Then u^* is continuous, and hence s^*, defined by u^* at points $x\in S$ with $|t(x)|\leq\varepsilon$ and by $s^*(x)=s(x)$ else, is transverse at c to L_o with $\#_c(s^*,L_o)=-1$. Using this procedure at every pole we replace s by an everywhere continuous section s^* transverse to L_o. At every zero a we have $\#_a(s^*,L_o)=\#_a(s,L_o)=1$, thus $\#(s^*,L_o)=k-n$.

<u>Corollary</u>: *The Euler number of* $\Sigma(n)$ *is* $-n$.

Proof: Using the presentation $\Sigma(n)=\{(z,w)\in\mathbb{C}^2\times\mathbb{P}: z_1w_2^n=z_2w_1^n\}$ there is the meromorphic section

$$z_1=\frac{w_1^n}{w_1^n-w_2^n}\ ,\quad z_2=\frac{w_2^n}{w_1^n-w_2^n}$$

with n simple poles and no zeros.

Finally we consider intersection numbers in the context of homology theory following the presentation in Dold's book, VIII,13 where a much more general situation is considered.

Let W be an oriented $(m+n)$-dimensional manifold, let $x\in H_m(W)$ and $y\in H_n(W)$ be two homology classes with integer coefficients. The *intersection number* $x\cdot y\in\mathbb{Z}$ is defined as follows: There are a compact neighbourhood retract $K\subset W$ and a class $y'\in H_n(K)$ such that $y'\mapsto y$ with $K\hookrightarrow W$. Let $\langle u,y'\rangle\in\mathbb{Z}$ for $u\in H^n(K)$ denote the pairing between cohomology and homology. Let $\mathcal{D}: H_m(W,W\setminus K)\cong H^n(K)$ denote the Poincaré duality as in Dold, VIII, 7.2. Then

$$x\cdot y=\langle\mathcal{D}j_*(x),y'\rangle\quad,$$

where $j:(W,\emptyset)\hookrightarrow(W,W\setminus K)$ denotes the embedding. This pairing is bilinear and symmetric up to sign,

$$x\cdot y=(-1)^{mn}y\cdot x\quad.$$

Let us return to the differentiable situation as described in the beginning of this section, let M and N be compact. The orientation of N determines a homology class $[N] \in H_n(N)$, hence by inclusion $N \hookrightarrow W$ also a class $[N] \in H_n(W)$.

The intersection number of differential topology equals the intersection number of homology theory,

(5) $$\#(f,N) = f_*([M]) \cdot [N] \quad .$$

Since textbooks either treat differential topology or homology theory but not the relations of both, a proof of (5) will be sketched: Let D be a neighbourhood of 0 in $\mathbb{R}^m$. A continuous map $\sigma\colon D \to W$ which is transverse to N with $\sigma^{-1}(N) = \{0\}$ is called a slice of N. The orientation of $\mathbb{R}^m$ determines a generator ι_m of $H_m(D, D \setminus \{0\}) \cong \mathbb{Z}$. Then

(6) $$\#_0(\sigma,N) = \langle \mathcal{D}\sigma_*(\iota_m), [N] \rangle$$

where $H_m(D,D\setminus\{0\}) \xrightarrow{\sigma_*} H_m(W,W\setminus N) \xrightarrow{\mathcal{D}} H^n(N)$ are isomorphisms. So far compare Dold, VIII, 11.1-7. where $\nu = \pm\sigma_*(\iota_m) \in H_m(W,W\setminus N)$ is called the transverse class of N in W. Let $f^{-1}(N) = \{x_1,..,x_r\} \subset M$. Choose mutually disjoint coordinate neighbourhoods M_i of x_i with oriented charts $h_i\colon (M_i,x_i) \to (D_i,0) \hookrightarrow (\mathbb{R}^m,0)$. Then $f \circ h_i^{-1}$ is a slice with

(7) $$\#_0(f \circ h_i^{-1},N) = \#_{x_i}(f,N) \quad .$$

The following maps induce isomorphisms of the m-th homology which is $\cong \mathbb{Z}$:

$$(M,\emptyset) \hookrightarrow (M,M\setminus\{x_i\}) \hookleftarrow (M_i,M_i\setminus\{x_i\}) \xrightarrow[\approx]{h_i} (D_i,D_i\setminus\{0\}) \quad .$$

The generator $[M] \in H_m(M)$ corresponds to the generator $\iota_m \in H_m(D_i,D_i\setminus\{0\})$. Let $[M_i]$ denote the corresponding element in $H_m(M_i,M_i\setminus\{x_i\})$. By (6) and (7)

$$\#_{x_i}(f,N) = \langle \mathcal{D}f_{i*}[M_i], [N] \rangle$$

for $f_i\colon (M_i,M_i\setminus\{x_i\}) \to (W,W\setminus N)$ induced by f. The summation over $i=1,..,r$ yields (5) because

(8) $$\sum_{i=1} f_{i*}([M_i]) = f_*([M])$$

in $H_m(W,W\setminus N)$ for $f\colon (M,\emptyset) \to (W,W\setminus N)$. For the proof of (8) the commutative diagram on the following page is considered. The horizontal arrows are induced by inclusions.

$$\begin{array}{ccccc} H_m(M) & \rightarrow & H_m(M,M\setminus f^{-1}(N)) & \xleftarrow{\sum_i j_{i*}} & \bigoplus_{i=1} H_m(M_i,M_i\setminus\{x_i\}) \\ & \searrow^{f_*} & \downarrow f_* & \swarrow_{\sum_i f_{i*}} & \\ & & H_m(W,W\setminus N) & & \end{array}$$

The elements $[M]\in H_m(M)$ and $([M_1],\ldots,[M_r])\in\bigoplus_i H_m(M_i,M_i\setminus\{x_i\})$ have the same image in $H_m(M,M\setminus f^{-1}(N))$.

§12 The Homology of Plumbed Surfaces

In order to calculate the homology of plumbed surfaces we use the results of the prec e ding section and the

Mayer-Vietoris-Sequence: *Let* X *and* Y *be open in the topological space* $X\cup Y$ *with inclusions according to the commutative diagram*

$$\begin{array}{ccc} X\cap Y & \xhookrightarrow{i} & X \\ j\downarrow & & \downarrow f \\ Y & \xhookrightarrow{g} & X\cup Y \end{array} .$$

There is a long exact sequence for the singular homology with field or integer coefficients

$$\ldots \rightarrow H_q(X\cap Y) \xrightarrow{\alpha} H_q(X)\oplus H_q(Y) \xrightarrow{\beta} H_q(X\cup Y) \xrightarrow{\partial_*} H_{q-1}(X\cap Y) \rightarrow \ldots$$

where $\alpha(z)=(i_*(z),j_*(z))$ *and* $\beta(x,y)=f_*(x)-g_*(y)$.

Theorem: *Let* X *be a complex surface which is plumbed according to the weighted tree* (T,b) *with* k *vertices.*

Then $H_o(X)=\mathbb{Z}$ *(i.e.* X *is path-connected)* , $H_q(X)=0$ *for* $0\neq q\neq 2$ *and* $H_2(X)$ *is free of rank* k .

A base of $H_2(X)$ *consists of the homology classes* $[E_1],\ldots,[E_k]$ *of the projective lines* E_i *which form the core* $E = \bigcup_i E_i$ *of* X . *With respect to this base the intersection form* $H_2(X)\times H_2(X) \to \mathbb{Z}$ *is given by the following* $k\times k$ *matrix* $S=(s_{ij})$: *The diagonal elements are* $s_{ii}=-b_i$. *Off the diagonal* $s_{ij}=0$ *if* $i \not- j$ *and* $s_{ij}=1$ *if* $i - j$.

Proof: By definition $X=X_1\cup\ldots\cup X_k$ is a union of tubular neighbourhoods X_i of the zero section E_i in $\Sigma(b_i)$. Each $E_i=\mathbb{P}_1\approx S^2$ is a deformation retract of X_i. This yields the desired homology for $k=1$. We proceed by induction and assume that $Y=X_1\cup\ldots\cup X_{k-1}$ has the desired homology. The intersection $Y\cap X_k$ is biholomorphic to the cartesian product $D\times D$ of two open disks $D\subset\mathbb{C}$. It can be contracted to a point, hence $H_q(Y\cap X_k)=0$ for $q>0$ and $=\mathbb{Z}$ for $q=0$. By induction hypothesis $H_q(Y) \oplus H_q(X_k) = \mathbb{Z} \oplus \mathbb{Z}$ for $q=0$, $= \mathbb{Z}^{k-1}\oplus\mathbb{Z}$ for $q=2$, and $=0$ else. The exact Mayer-Vietoris sequence

$$\ldots \to H_q(Y\cap X_k) \to H_q(Y) \oplus H_q(X_k) \to H_q(Y\cup X_k) \to H_{q-1}(Y\cap X_k) \to \ldots$$

yields the desired homology of $X=Y\cup X_k$.

The intersection matrix (s_{ij}) is obtained by means of the results of §11: The self-intersection number s_{ii} of E_i in X depends only on the neighbourhood X_i . Therefore $s_{ii}=$ self-intersection number of the zero section of $\Sigma(b_i)$, which is $=-b_i$ by the corollary at the end of §11. If $i - j$, the projective lines E_i and E_j intersect transversely in one point with local intersection number $+1$, hence $s_{ij}=1$. If $i \not- j$, they do not intersect, hence $s_{ij}=0$.

The reader may now establish the intersection matrices of the resolution of the quotient singularities $\mathbb{C}^2/G$ using §8 , theorem 2 and 3. By direct calculation he will state that the matrices obtained are negative definite. There is a general result behind this: The intersection matrix of the resolution of an irreducible surface singularity is always negative definite, see H. Laufer: Normal two-dimensional Singularities, Annals of Mathematics Studies 71, Princeton N.J. 1971, Theorem 4.4 on page 49.

Next the homology of the complement $X \setminus E$ is calculated. As above $X = X_1 \cup \ldots \cup X_k$ is a plumbed surface with core $E = E_1 \cup \ldots \cup E_k$. The inclusion $E \hookrightarrow X$ induces isomorphisms of the homology. This follows from the preceding theorem. Part of the exact homology sequence of the pair $(X, X \setminus E)$ is as follows:

$$H_3(X, X\setminus E) \to H_2(X \setminus E) \to H_2(X) \xrightarrow{j_*} H_2(X, X\setminus E) \to H_1(X\setminus E) \to H_1(X) .$$

Here $H_3(X, X\setminus E) \cong H^1(E)$ by Poincaré-Lefschetz duality and $H^1(E) = H^1(X) = H_1(X) = 0$. Hence

$$H_2(X\setminus E) = \text{kernel of } j_* , \quad H_1(X\setminus E) = \text{cokernel of } j_* .$$

Now $H_2(X, X\setminus E) = H^2(E)$ by duality, $H^2(E) \cong H^2(X)$ by inclusion and $H^2(X) = \mathrm{Hom}(H_2(X), \mathbb{Z})$ by the universal coefficient theorem. Thus j_* corresponds to the homomorphism $\iota : H_2(X) \to \mathrm{Hom}(H_2(X), \mathbb{Z})$ which is given by $\langle \iota(x), y \rangle = x \cdot y$. Here $\langle u, y \rangle \in \mathbb{Z}$ denotes the value of $u \in \mathrm{Hom}(A, \mathbb{Z})$ at $y \in A$ and $x \cdot y$ denotes the intersection pairing. According to the preceding theorem we have the base $[E_1], \ldots, [E_k]$ of the free module $H_2(X)$. Let $\varepsilon_1, \ldots, \varepsilon_k$ denote the dual base of $\mathrm{Hom}(H_2(X), \mathbb{Z})$, i.e. $\langle \varepsilon_i, [E_j] \rangle = \delta_{ij}$ (Kronecker symbol). With respect to these bases ι is given by the intersection matrix $S = (s_{ij})$. Thus

<u>Theorem 2</u>: *Let* $S : \mathbb{Z}^k \to \mathbb{Z}^k$ *denote the homomorphism given by the intersection matrix. Then* $H_2(X\setminus E)$ *is isomorphic to the kernel of* S *and* $H_1(X\setminus E)$ *is isomorphic to the cokernel of* S .

By elementary line and column operations S is brought into diagonal form

$$D = \begin{bmatrix} d_1 & & & & \\ & \ddots & & 0 & \\ & & d_r & & \\ & 0 & & 0 & \\ & & & & \ddots \\ & & & & & 0 \end{bmatrix}$$

with positive integers d_i such that d_i divides d_{i+1}. There are automorphisms Φ and Ψ of $\mathbb{Z}^k$ such that $\Psi \circ S = D \circ \Phi$. Hence up to isomorphism $H_2(X \setminus E)$ = kernel of D and $H_1(X \setminus E)$ = cokernel of D.

<u>Corollary</u>: $H_1(X \setminus E) = \mathbb{Z}/d_1\mathbb{Z} + \ldots + \mathbb{Z}/d_r\mathbb{Z} + \mathbb{Z}^{k-r}$

$H_2(X \setminus E) = \mathbb{Z}^{k-r}$.

For a finite subgroup $G < SL(2,\mathbb{C})$ let $\sigma: X \to \mathbb{C}^2/G$ be the resolution of the quotient singularity as described in §10, theorem 2 and 3. Let $E \subset X$ be the exceptional set. Then $r=k$, i.e. $H_2(X \setminus E) = 0$ and $H_1(X \setminus E)$ is finite:

type of G	$H_1(X \setminus E) \cong$
cyclic of order n	$\mathbb{Z}/n\mathbb{Z}$
binary q-dihedral	$\mathbb{Z}/2\mathbb{Z} \oplus \mathbb{Z}/2\mathbb{Z}$ for even q, $\mathbb{Z}/4\mathbb{Z}$ for odd q
binary tetrahedral	$\mathbb{Z}/3\mathbb{Z}$
binary octahedral	$\mathbb{Z}/2\mathbb{Z}$
binary icosahedral	0

The neccessary calculation are left to the reader but see §14 for another proof.

§13 The Fundamental Group of a Plumbed Surface Minus its Core

We shall use an extended version of the Seifert-van Kampen Theorem for the fundamental group of a union: Let $Y=Y_1\cup\ldots\cup Y_k$ be a finite union of open path-connected subsets. Let each intersection $Y_i\cap Y_j$ also be path-connected or empty. Let T be the following graph: It has one vertex i for each Y_i. Two vertices i and j are connected by an edge $(i—j)$ if and only if $Y_i\cap Y_j$ is not empty. There is a continuous mapping $\tau: T \to Y$ such that each vertex i is mapped into the corresponding Y_i and such that the edge $i—j$ is mapped into $Y_i\cup Y_j$. *From now on this graph* T *is assumed to be a tree.*

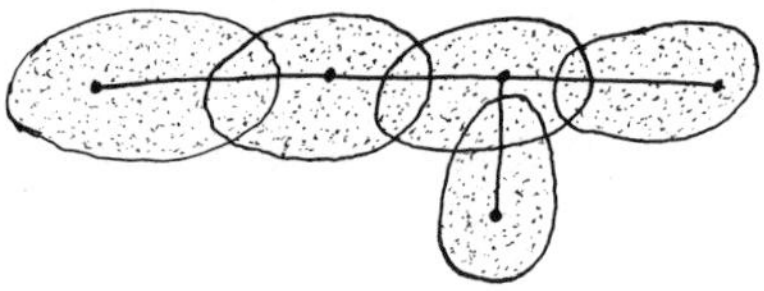

<u>Convention on the base points</u>: The base points are always assumed to lie in the image $\tau(T)$. Any two base points x and y are connected by a path s which factors through τ. Since T is contractible (being an tree), the isomorphism $s_*: \pi(Y,x) = \pi(Y,y)$ does not depend on s. We shall identify $\pi(Y,x)=\pi(Y,y)$ and simply write $\pi(Y)$.

<u>Seifert-van Kampen Theorem (extended version)</u>: *Let each* $\pi(Y_i)$ be *presented by a set of generators* G_i *and a set of relations* R_i. *If* $i—j$, *let* $Y_i \xleftarrow{f} Y_i\cap Y_j \xrightarrow{g} Y_j$ *be the inclusions. Let* G_{ij} *be* a set of *generators* γ *of* $\pi(Y_i\cap Y_j)$. *Let* $\gamma_i =: f_*(\gamma)$ and $\gamma_j =: g_*(\gamma)$ *be expressed by the elements of* G_i *and* G_j *respectively. Let* $R_{ij} =: \{\gamma_i\gamma_j^{-1}: \gamma\in G_{ij}\}$.
Then $\pi(Y)$ *is presented by the set of generators*

$\coprod_{\{vertices\}} G_i = G_1 \amalg \ldots \amalg G_k$ *(disjoint union) and the set of relations* $\bigcup_{\{vertices\}} R_i \cup \bigcup_{\{edges\}} R_{ij}$.

The classical theorem is the case $k=2$, see e.g. Massey Chap.IV. From this the extended version is easily proved by induction on k following the way in which a tree is built up.

Let (T,n) be a tree with k vertices weighted by the integers $n_1,\ldots,n_k$. Let the vertices be ordered in some way. Let X be a complex surface plumbed according to (T,n) with core E .

<u>Theorem</u>: *The fundamental group of* $X\setminus E$ *is presented by a set of* k *generators* $\{v_1,\ldots,v_k\}$ *and the following relations:*

If i—j*, then* v_i *and* v_j *commute* .

$$\prod_{j—1} v_j = v_1^{n_1} \quad , \ldots , \quad \prod_{j—k} v_j = v_k^{n_k} \quad .$$

Here $\prod_{j—i}$ means the product over all j's which are connected to the fixed i . The factors are ordered according to increasing j's .

Proof: If i—j, we use the holomorphic chart (z,w) of $X_i \cap X_j$ given in § 10 . We fix a base point $b_{ij} \in (X_i \cap X_j)\setminus E$ with coordinates (z_o,w_o) . The loops

(1) $v_{ij}(t) =: (z_o e^{\sqrt{-1}t}, w_o)$ and $v_{ji}(t) =: (z_o, w_o e^{\sqrt{-1}t})$ for

$0 \le t \le 2\pi$ are contained in $X_i \cap X_j$, see figure 1 .

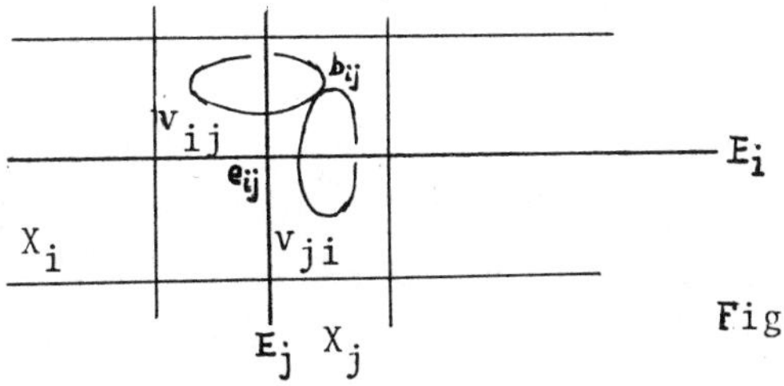

Figure 1

Denote by e_{ij} the point where E_i and E_j intersect. Let $\dot{E}_i = E_i \setminus \{e_{ij} : j — i\}$. Let $p: X_i \to E_i$ be the bundle projection. Then $u_j = p \circ v_{ij}$ is a simple loop in $\dot{E}_i$ around e_{ij} from and to $a_{ij} = p(b_{ij})$. We choose a base point a_i and paths s_j in $\dot{E}_i$ from a_i to a_{ij} such that for the simple loops $u_{ij} = s_j u_j s_j^{-1}$ in $\dot{E}_i$ we have

$$\prod_{j—i} u_{ij} = 1 \quad \text{in} \quad \pi(\dot{E}_i, a_i) \quad .$$

Figure 2 shows $\dot{E}_i$ for three points $e_{i\nu}$. By stereographic projection from one of these points, $\infty = e_{ih}$, $\dot{E}_i$ becomes a plane punctered at two points.

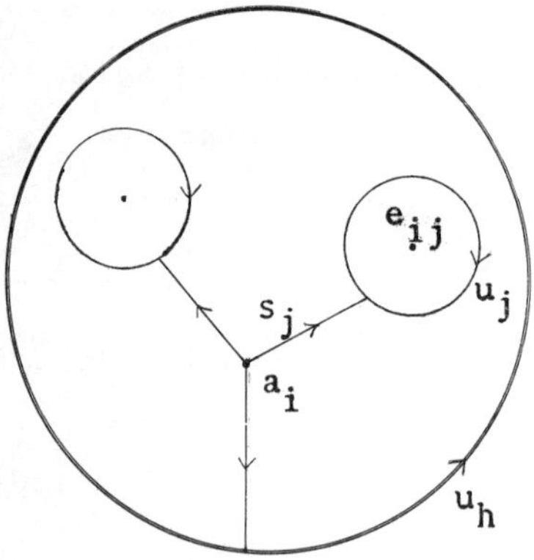

Figure 2

We lift the base point a_i to $b_i \in X_i \setminus E$ and lift the paths s_j to paths s_{ij} in $X_i \setminus E$ such that s_{ij} begins in b_i and ends in b_{ij}. Here "lifting" means $p(b_i) = a_i$ and $p \circ s_{ij} = s_j$. The tree T is mapped continuously into $X \setminus E$ in the following way: Each vertex i is mapped to b_i. Half the edge from i to j is mapped as path s_{ij} and the other half from j to i as s_{ji} so that the mid-edge point becomes b_{ij}, see figure 3. This mapping $T \to E$ is used for the convention on the base points.

Let us now fix i. Let $h \neq i$ be the smallest vertex connected to i. In E_i we choose projective coordinates $[z_o : z_1]$ such that $a_i = [1:0]$ and $e_{ih} = [0:1]$. There is a trivialization $(z', w'): X_i \setminus E_h \hookrightarrow \mathbf{C}^2$ with $z' = (z_1/z_o) \cdot p$ and with w' being a linear fibre coordinate. This trivialization

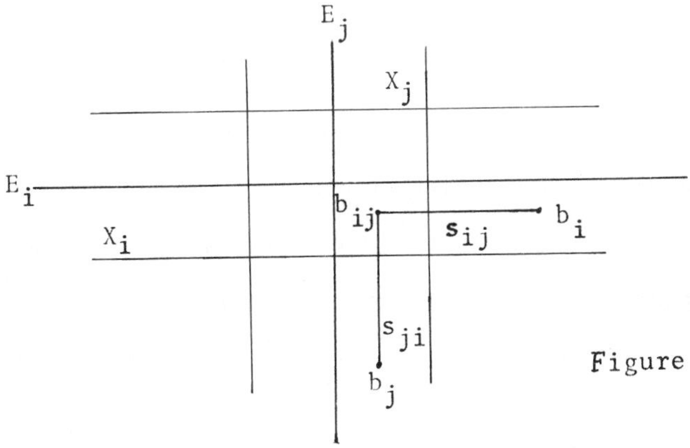

Figure 3

shows that $X_i \setminus E$ and $\dot{E}_i \times \dot{D}$ are homeomorphic. Here D denotes a disk in $\mathbb{C}$ around 0 and $\dot{D} = D \setminus \{0\}$. This implies for the fundamental groups

$$\pi(X_i \setminus E) \cong \pi(\dot{E}_i \times \dot{D}_i) = \pi(\dot{E}_i) \times \pi(\dot{\mathbb{C}}) .$$

Here $\pi(\dot{E}_i)$ is the free group generated by $\{u_{ij} : j - i$ but $j \neq h\}$ and $\pi(\dot{\mathbb{C}})$ is infinite cyclic generated by a simple loop c around 0. Let $\tilde{v}_{ij}$ be a simple loop around E_j in $X_i \setminus E$, i.e. $p \circ \tilde{v}_{ij} = u_{ij}$ in $\pi(\dot{E}_i)$ and $w' \circ \tilde{v}_{ij} = 1$ in $\pi(\dot{\mathbb{C}})$. Let v_i be a simple loop around E_i in $X_i \setminus E$, i.e. $p \circ v_i = 1$ in $\pi(\dot{E}_i)$ and $w' \circ v_i = c$ in $\pi(\dot{\mathbb{C}})$. Thus $\pi(X_i \setminus E)$ is presented by the set of generators

(2) $$\{\tilde{v}_{ij} : j - i \text{ but } j \neq h\} \cup \{v_i\}$$

and the relations

(3) $$\tilde{v}_{ij} \text{ and } v_i \text{ commute} .$$

(4) *The loop* v_{ij} *defined by* (1) *is a simple loop in* $X_i \setminus E$ *around* E_j *provided* $j \neq h$.

Proof: The coordinates (z,w) which have been used in order to define v_{ij} are transformed into the coordinates (z',w')

$$z' = f(z) , \quad w' = g(z)\, w$$

with holomorphic functions such that $f(0) = z'(e_{ij}) =: \zeta_j$,

$f'(0)\neq 0$ and $g(0)\neq 0$. The winding number of $z'\circ v_{ij}(t)=f(z_o e^{\sqrt{-1}t})$ around ζ_ν is the order of $f-\zeta_\nu$ at 0 . This is 1 if $\nu=j$ and 0 else. Thus $p\circ v_{ij}=u_j$ in $\pi(\dot{E}_i)$. Similarly the winding number of $w'\circ v_{ij}(t)=g(z\, e^{\sqrt{-1}t})\cdot w_o$ around 0 is the order of g at 0 which is 0 so that $w'\circ v_{ij}=1$ in $\pi(\dot{\mathbb{C}})$. *We may therefore replace* $\tilde{v}_{ij}$ *by* v_{ij} *in* (2).

Next we want to compute v_{ih} in terms of the v_{ij} with $j\neq h$ and v_i . There is a holomorphic chart (z,w) defined on $X_i\smallsetminus p^{-1}(a_i)$ with $z=(z_o/z_1)\circ p$ and with w being a linear fibre coordinate, such that

(5) $$z'=1/z \quad\text{and}\quad w'=z^{n_i}\cdot w \quad ,$$

see § 4(4) . Though (z,w) may not be the coordinates which have been used in order to define v_{ih} by (1) we may assume $v_{ih}(t)=(u_o e^{\sqrt{-1}t},v_o)$ for (z,w) occuring in (5) using the same argument as for (4). Now $z'\circ v_{ih}(t)=u_o^{-1}e^{-\sqrt{-1}t}$ so that $p\circ v_{ih}$ is the big circle of figure 2 and hence $p\circ v_{ih}=(\prod'_{j-i} u_j)^{-1}$ with Π'=product over the j's different from h . Furthermore $w'\circ v_{ih}(t)=u_o^{n_i}v_o e^{\sqrt{-1}n_i t}$. This has winding number n_i around 0 in $\mathbb{C}$, thus $w'\circ v_{ih}=c^{n_i}$ in $\pi(\mathbb{C})$. Combining these results we obtain $v_{ih}=v_i^{n_i}(\prod'_{j-i} v_{ij})^{-1}$. Thus the presentation (2)(3) of $\pi(X_i\smallsetminus E)$ is equivalent to the presentation with generators

(6) $$\{v_{ij}\colon j-i\}\cup\{v_i\}\ ,\ \text{i fixed}\ ,$$

and relations

(7a) v_i and v_{ij} commute ,

(7b) $$\prod_{j-i} v_{ij}=v_i^{n_i} \quad .$$

The presentation of $\pi(X \setminus E)$ is now obtained using the extended version of the Seifert-van Kampen theorem applied to $X \setminus E = (X_1 \setminus E) \cup \ldots \cup (X_k \setminus E)$: To the generators (6) and the relations (7) of $\pi(X_i \setminus E)$ for $i=1,\ldots,k$ further relations are added which come from the inclusions $X_i \setminus E \xhookleftarrow{f} (X_i \cap X_j) \setminus E \xhookrightarrow{g} X_j \setminus E$ for $i — j$: Since $(X_i \cap X_j) \setminus E$ is homeomorphic to $\dot{D} \times \dot{D}$, see §10(3), $\pi((X_i \cap X_j) \setminus E)$ is generated by v_{ij} and v_{ji}, see figure 1. Now $f \circ v_{ij} = v_{ij}$ is a generator of $\pi(X_i \setminus E)$ occuring in (6). Furthermore $g \circ v_{ij} = v_j \in \pi(X_j \setminus E)$ because it is a simple loop within a fibre of $p\colon X_j \setminus E_j \to E_j$. Therefore the additional relations are

(8) $$v_{ij} = v_j \quad \text{for } i — j\,.$$

The presentation given in the theorem follows easily from (6), (7), and (8).

§ 14 Groups Determined by a Weighted Tree

The theorem in the preceding §13 tells us how to obtain a presentation of the fundamental group of a plumbed surface minus its core from the tree defining the plumbing. We consider some examples:

Linear trees:

(1)

The group is presented by k generators $v_1,\ldots,v_k$ and the relations

(2) $\quad v_2=v_1^{n_1}\ ,\ v_1v_3=v_2^{n_2}\ ,\ \ldots,\ v_{k-1}=v_k^{n_k}$

We calculate the "Euclidean" algorithm

$$q_o=1$$
$$q_1=n_1q_o$$
$$q_2=n_2q_1-q_o$$
$$\vdots$$
$$q_k=n_kq_{k-1}-q_{k-2}$$

The relations (2) imply $v_{i+1}=v_1^{q_i}$ for $i=0,\ldots,k-1$ and and $1=v_1^{q_k}$, i.e. *The group is cyclic of order* q_k.

When we attach the generators v_i to the vertices i of the linear tree, it looks as follows $(v=v_1)$:

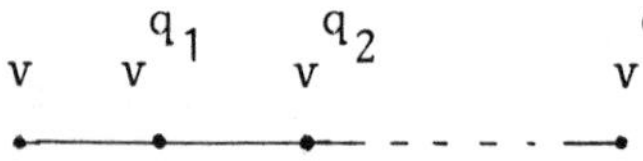

and $v^{q_k}=1$.

If all weights $n_i=2$, we obtain particularly

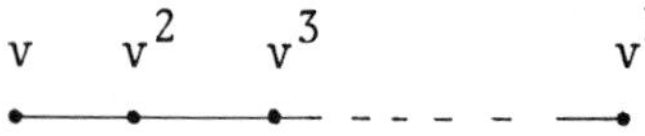

and $v^{k+1}=1$.

The trees $T_{p,q,r}$ look as follows

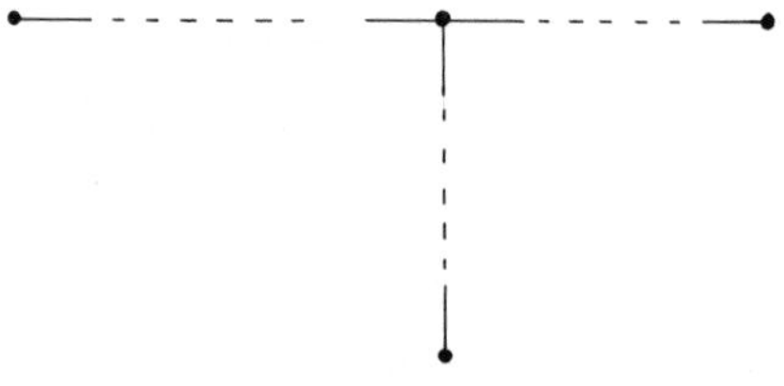

The three branches have $p,q,$ and r vertices respectively when the triple vertex is included each time. The total number of vertices is $p+q+r-2$. Let all weights = 2. Using the result on linear graphs we obtain the following picture of generators attached to the vertices

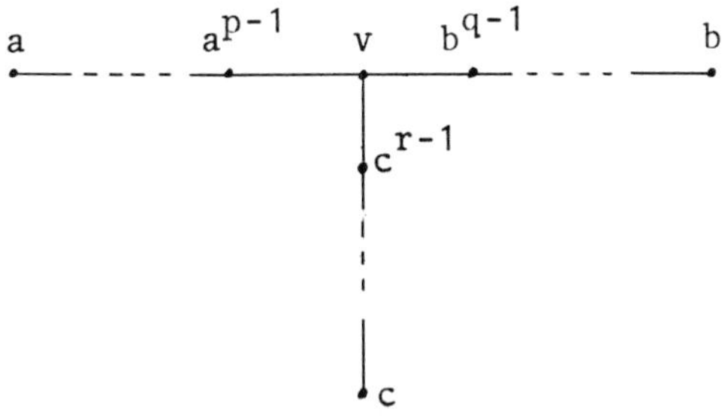

and the relations $v=a^p=b^q=c^r$. The relation at the triple vertex is $a^{p-1}c^{r-1}b^{q-1}=v^2$. Since $v^2=a^pb^q$ this gives $c^r=abc$. Thus:

The group determined by the tree $T_{p,q,r}$ *weighted by* 2 *at every vertex is presented by three generators* a,b,c *and the relations*

(3) $$a^p=b^q=c^r=abc \quad .$$

Obviously the element $v=abc$ belongs to the center of the group.

By making the fundamental group *abelian* the first homology group with integer coefficients is obtained. When the tree is linear the group is already abelian whence

(4) $$H_1(X\setminus E) = \mathbb{Z}/q_k\mathbb{Z}$$

for a surface X plumbed according to (1). For the trees $T_{p,q,r}$ the abelianized group, written additively, is given by three generators a,b,c and the relations $pa=qb=rc=a+b+c$. We choose three relations and form their coefficient matrix

$$\begin{pmatrix} p & -q & 0 \\ 0 & q & -r \\ 1-p & 1 & 1 \end{pmatrix} \quad .$$

By adding suitable multiples of the last row and column this is transformed into

$$\begin{pmatrix} p & -q & 0 \\ r(1-p) & r+q & 0 \\ 0 & 0 & 1 \end{pmatrix} .$$

The absolute value of the determinant of the matrix, i.e

(5) $$pqr\left|\frac{1}{p}+\frac{1}{q}+\frac{1}{r}-1\right|$$

is the order of the abelianized group. (Determinant=0 means infinite order. If without loss of generality $p \leq q \leq r$ is assumed, this can only occur in three cases $(p,q,r)=(2,3,6)$, $(2,4,4)$, and $(3,3,3)$.) In order to determine the structure of the abelianized group further row and colums operations must be applied in order to obtain a matrix of the form

$$\begin{pmatrix} d_1 & 0 & 0 \\ 0 & d_2 & 0 \\ 0 & 0 & 1 \end{pmatrix} \quad \text{with} \quad d_2 \text{ divides } d_1, \text{ possibly } d_1=0 .$$

The abelianized group is $\mathbb{Z}/d_1\mathbb{Z} \oplus \mathbb{Z}/d_2\mathbb{Z}$.

We are especially interested in the trees $D_{r+2}=:T_{2,2,r}$ for $r=2,3,...$ and $E_{r+3}=:T_{2,3,r}$ for $r=3,4$, and 5 because they describe the resolution of non-cyclic quotient singularities, see §25, theorem 2. Using the matrix transformations described above this yields the following abelianized groups

D_k	D_k	E_6	E_7	E_8
k even	k odd			
$\mathbb{Z}/2\mathbb{Z} \oplus \mathbb{Z}/2\mathbb{Z}$	$\mathbb{Z}/4\mathbb{Z}$	$\mathbb{Z}/3\mathbb{Z}$	$\mathbb{Z}/2\mathbb{Z}$	$\{0\}$

These results as well as the result (4) are in accordance with the table at the end of §12.

Let us return to the *non-abelianized* groups for a

Final Observation: Let G denote a cyclic subgroup $C_{n,q} <$ $GL(2,\mathbb{C})$ or an arbitrary finite subgroup of $SL(2,\mathbb{C})$. Let (T,b) denote the weighted graph which is assigned to the quotient singularity $\mathbb{C}^2/G$, see §10, Theorems 1 and 2. We compare the presentation of the group $\Gamma(T,b)$ which is determined by (T,b) with the presentation of G in II§5 and observe the isomorphism

$$G \cong \Gamma(T,b)$$

by checking the different groups under consideration. An explanation will be given in the following section.

§ 15 Topological Invariants

In §6 and §9 we constructed a resolution $f: X \to \mathbb{C}^2/G$ of the quotient singularity of every finite subgroup $G < SL(2,\mathbb{C})$. A neighbourhood Y of the exceptional set $E=f^{-1}(\pi(0))$ turned out to be a plumbed surface (§10). Using this we calculated some topological invariants of Y and $Y \setminus E$ (§12 and §13). *Actually these are invariants of the singularity* $(\mathbb{C}^2/G,\pi(0))$. But a priori this is not clear because the resolution f and the neighbourhood Y could have been chosen otherwise, resulting in other invariants. We shall see that both the resolutions and the neighbourhoods, we chose, belong to distinguished classes which yield isomorphic invariants.

The distinguished resolutions are the minimal ones, compare Laufer's book: Let V be a 2-dimensional complex space which has a normal singularity at c . A resolution $f: \tilde{V} \to V$ at c is called *minimal* if for every other resolution $g: X \to V$ at c there is a unique holomorphic map $\varphi: X \to \tilde{V}$ with $f\circ\varphi = g$. A general argument shows that the minimal resolution is unique.

Criterion for Minimal Resolutions: *The resolution* $f: \tilde{V} \to V$ *at* c *is minimal if the exceptional set* $f^{-1}(c)=E=E_1\cup\ldots\cup E_k$ *is the union of regular complex curves* E_i *such that no projective line* E_j *with self-intersection number* $[E_j]\cdot[E_j]=-1$ *occurs.*

Blowing up the origin of $\mathbb{C}^2$ is a typical resolution which is not minimal. With the resolutions of quotient singularities, which we constructed, all E_j are projective lines, but their self-intersection numbers are always $-b_j<-1$, thus we obtain

Proposition 1: *The resolutions of quotient singularities* $\mathbb{C}^2/G$, *which have been constructed in §6 and §9, are uniquely determined as* the *minimal resolutions.*

The distinguished neighbourhoods are the "homotopy small" ones. Let A be a closed subspace of the topological space X. A continuous map $r: X \to A$ is called a *retraction* if $r(a)=a$ for every $a\in A$. It is called a *deformation retraction* if in addition there is a continuous map (deformation) $h: X\times[0,1]\to X$ with $h(x,0)=r(x)$ and $h(x,1)$ for every $x\in X$. The subspace A is called a (*deformation*) *retract* of X if there is a (deformation) retraction $r: X\to A$.

Lemma 2: *Let* $B\subset A\subset X$. *If* A *is a retract of* X *and if* B *is a deformation retract of* X, *then* B *is a deformation retract of* A.

Proof: Let $r: X\to A$ be the retraction and let $h: X\times[0,1]\to X$ be the deformation of X into B. Then $r\circ h|A\times[0,1]\to A$ is a deformation of A into B.

A neighbourhood U of A is called *homotopy small* if there are arbitrary small neighbourhoods U_o of A such that U_o is a deformation retract of U and $U_o^-=U_o\setminus A$ is a deformation retract of $U^-=U\setminus A$. *Then* U_o *is also homotopy small* because of lemma 2.

Proposition 3: *If* U *and* V *are homotopy small neighbourhoods of* A, *then* U *and* V *as well as* U^- *and* V^- *have the same homotopy type.*

Proof: We may assume that $V \subset U$. There are neighbourhoods U_0 and V_0 with $U \supset V \supset U_0 \supset V_0 \supset A$ and with the following deformation retracts: U_0 of U , U_0^- of U^- , V_0 of V , V_0^- of U_0^- ,further V_0 of U_0 , and V_0^- of U_0^- .

In order to recognize homotopy small neighbourhoods we use *controlled deformations*. Let A be a closed subspace of X and let $h: X \times [0,1] \to X$ be a deformation onto A . A continuous function $\rho: X \to \mathbb{R}$ is said to control h if

$\rho(x) \geq 0$ for every $x \in X$ and $\rho(x)=0$ if and only if $x \in A$.

$\rho \circ h(x,t) = t\rho(x)$ for every $x \in X \times [0,1]$.

Then $h(x,t) \in A$ if and only if $(x,t) \in X \times \{0\} \cup A \times [0,1]$. The closed set A is also a controlled deformation retract of every

$$X_\varepsilon = \{x \in X: \rho(x) \leq \varepsilon\}, \quad \varepsilon > 0 .$$

The map

$r_\varepsilon: X \to X_\varepsilon$, $r_\varepsilon(x)=x$ if $\rho(x) \leq \varepsilon$, and $r_\varepsilon(x)=h(x,\varepsilon/\rho(x))$ if $\rho(x) \geq \varepsilon$

is a deformation retraction, and so is the restriction r_ε: $X \setminus A \to X_\varepsilon \setminus A$. The corresponding deformation is $(x,t) \mapsto r_{\varepsilon+t-\varepsilon t}(x)$.

If X is locally compact and A is compact, the set $\{X_\varepsilon\}_{\varepsilon>0}$ is a neighbourhood base of A in X . We obtain

Proposition 4: *If* A *is a compact controlled deformation retract of the locally compact space* X *, then* X *is a homotopy small neighbourhood of* A .

Proposition 5: *Let* $f: Y \to X$ *be a modification at* $a \in X$ *, and assume that* $\{a\}$ *is a deformation retract of* X *, controlled by* ρ *. Then* Y *and every* $Y_\varepsilon = \{y \in Y: \rho f(y) \leq \varepsilon\}$ *are homotopy small neighbourhoods of the exceptional set* $E=f^{-1}(a)$.

Proof: The set $\{Y_\varepsilon\}_{\varepsilon>0}$ is a neighbourhood base of E . It suffices to show that Y_ε and $Y_\varepsilon \setminus E$ are deformation retracts of Y and $Y \setminus E$ respectively. We use r_s as defined above, and define the deformation $h: Y \times [0,1] \to Y$ of Y onto Y_ε by $h(y,t)=f^{-1} r_{\varepsilon+t-\varepsilon t} f(y)$ for $y \in Y \setminus E$ and $h(y,t)=y$ for $y \in Y_\varepsilon$.

<u>Proposition 6</u>: *The core* E *of a plumbed complex surface* X *is a controlled deformation retract of* X .

Proof: Let $X=X_1\cup\ldots\cup X_k$ with core $E=E_1\cup\ldots\cup E_k$, compare §10. Each X_i is a tubular neighbourhood of the zero section of a complex line bundle over E_i . Let $p_i: X_i \to E_i$ denote the bundle projection. For $i \neq j$ the intersection $X_i \cap X_j$ is either empty or isomorphic to $\{(z,w) \in \mathbb{C}^2 : |z|<1,\ |w|<1\}$. We define a retraction $r: X \to E$ as follows: If $x \in X_i, \notin X_j$ for $j \neq i$, then $r(x)=p_i(x) \in E_i$. If $x \in X_i \cap X_j$ for $j \neq i$ the coordinates (z,w) are used. For $0 \leq \beta \leq \alpha < 1$ let $\tau(\alpha,\beta) = (\alpha-\beta)/(1-\beta)$, see figure 1. Then define

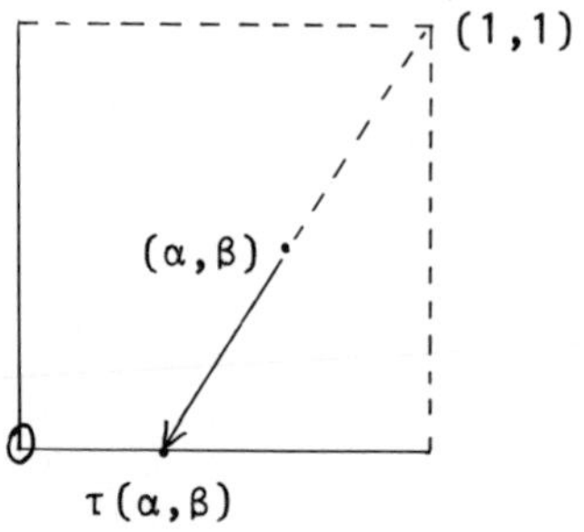

Figure 1

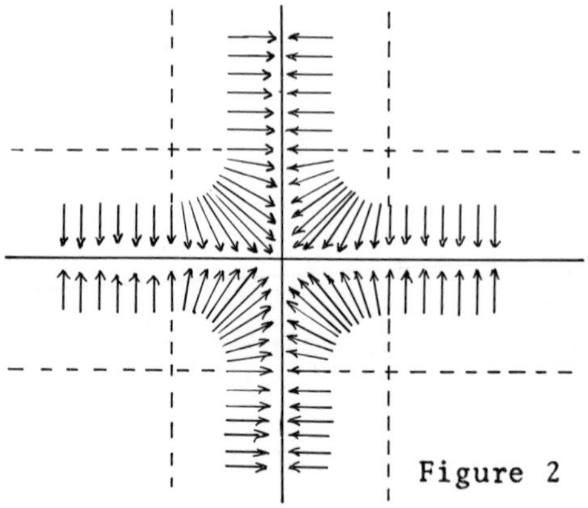
Figure 2

$$r(z,w)=(\tau(|z|,|w|)\ z/|z|,0) \quad \text{if} \quad 0\leq|w|\leq|z|<1$$
$$r(z,w)=(0,\ \tau(|w|,|z|)\ w/|w|) \quad \text{if} \quad 0\leq|z|\leq|w|<1$$

Figure 2 shows a picture of r in real dimensions.

This retraction r is actually a deformation retraction. We define the corresponding deformation $h: X\times[0,1] \to X$ as follows: If $x \in X_i, \notin X_j$ for $j \neq i$, then $h(x,t)=tx$, using the linear structures of the fibres of the bundle projection $p_i: X_i \to E_i$. If $x \in X_i \cap X_j$ for $i \neq j$, then $h(x,t)=tx+(1-t)r(x)$, using the linear structure given by the coordinates (z,w) on $X_i \cap X_j$. We observe that $r\circ h(x,t)=r(x)$.

In order to define a control function $\rho: X \to \mathbb{R}$ we choose norms $\rho_i: X_i \to \mathbb{R}$ on the line bundles $p_i: X_i \to E_i$ such that for every $i \neq j$ and every $x \in X_i \cap X_j$ with coordinates (z,w) we have $\rho_i(x)=|z|$ and $\rho_j(x)=|w|$. (This is always possible, using a partition of unity). Then the following $\rho: X \to \mathbb{R}$ is continuous and controls the deformation h:

$\rho(x)=\rho_i(x)$ if $x \in X_i, \notin X_j$ for $j \neq i$
$\rho(x)=\|x-\tau(x)\|$ if x $X_i \cap X_j$, where $\|..\|$ denotes the Hermitean norm given by the coordinates (z,w).

Let $G < GL(m,\mathbb{C})$ be a finite subgroup, let $p: \mathbb{C}^m \to V=\mathbb{C}^m/G$ denote the orbit projection with $c=p(0)$. Then $\{c\}$ is a deformation retract of V, controlled by $\rho(p(z))=\|z\|$, where $\|..\|$ is a G-invariant norm. Such norms exist by II§2(a). For the cyclic subgroup $C_{n,q} < GL(2,\mathbb{C})$, and for arbitrary finite subgroups $G < SL(2,,\mathbb{C})$ we constructed minimal resolutions $f: X \to V$ at c in §6 and §9. We apply proposition 5 and obtain:

(1) X *and every* $X_\varepsilon=\{x\in X: \rho f(x) \leq \varepsilon\}$ *are homotopy small neighbourhoods of the exceptional set* $E=f^{-1}(c)$.

There is a neighbourhood Y of E, which is a complex surface plumbed according to the tree (T,b) of the resolution f. By proposition 6, E is a deformation retract of Y, controlled by some ρ^*. Then

(2) Y *and every* $Y_\delta=\{x \in Y: \rho^*(x) \leq \delta\}$ *are homotopy small neighbourhoods of* E.

Thus by proposition 3, we obtain the following homotopy equivalences:

(3a) $X \simeq X_\varepsilon \simeq Y \simeq Y_\delta$ (3b) $X^- \simeq X_\varepsilon^- \simeq Y^- \simeq Y_\delta^-$ $(X^-=X-E$ etc$)$

The proof of proposition 3 even yields a bit more: There are α,β,γ such that $X \supseteq Y \supseteq X_\alpha \supseteq Y_\beta \supseteq X_\gamma$ with Y_β being a deformation retract of X_α and X_γ being a deformation retract of Y_β, similarly for X_α^-, Y_β^- and X_γ^-.

Since X is the unique minimal resolution (Proposition 1) the homotopy types of Y and of $Y \setminus E$ are invariants of the singularity (V,c), in particular

(4) *The homology of* Y *and of* $Y\setminus E$, *the intersection form on* $H_2(Y)$ *and the fundamental group* $\pi(Y\setminus E)$ *are invariants of the singularity* (V,c).

Let $p: \mathbb{C}^2 \to V$ dentoe the orbit projection of a finite subgroup $G<U(2)$ acting on $\mathbb{C}^2$, compare p.66. The radial deformation retration of the punctured unit ball $\{z \in \mathbb{C}^2: 0<\|z\| \leq 1\}$ onto the boundary sphere $S^3=\{z \in \mathbb{C}^2: \|z\|=1\}$ is G-invariant, and so induces a deformation retraction of $V_1=\{p(z) \in V: \|z\| \leq 1\}$ onto $\partial V_1=\{p(z) \in V: \|z\|=1\}=S^3/G$. The three-manifolds S^3/G have been studies in II§4 for the finite subgroups $G<SU(2)$. They are called *lens spaces* $L_{n,q}=S^3/C_{n,q}$ for the cyclic subgroups $C_{n,q}<U(2)$. The result of the theorem in II§4 about the fundamental group

(5) $$\pi(S^3/G) \cong G$$

holds also true for $G=C_{n,q}$. Since $f: X\setminus E \to V\setminus\{c\}$ is biholomophic, the boundary $\partial X_1=\{x\in X: \rho f(x)=1\}$ of the compact three-manifold X_1 is a deformation retract of $X_1\setminus E \approx V_1\setminus\{c\}$ and is homeomorphic to $V_1=S^3/G$. Therefore by (3b),

(6) S^3/G *and* $Y\setminus E$ *are homotopy-equivalent.*

In particular, they have isomorphic fundamental groups. The fundamental group of $Y\setminus E$ is the group determined by the weighted tree (T,b). This together with (5) explains the final observation of the preceding §14.

Mumford showed that for any normal surface singularity (V,c) the fundamental group $\pi(V\setminus\{c\})$ is trivial if and only if V is regular at c.

The fundamental group $\pi(S^3/G)\cong G$ distinguishes the different S^3/G up to homeomorphy for the finite subgroups $G < SL(2,\mathbb{C})$ but not for the cyclic subgroups $G=C_{n,q} < GL(2,\mathbb{C})$: Two lens spaces $L_{n,q}$ and $L_{n,p}$ are homeomorphic if and only if $qp\equiv\pm 1 \mod n$. The proof requires the refined notion of 'simple homotopy type'. A good reference is M.M. Cohen's book.

CHAPTER V

THE HIERARCHY OF SIMPLE SINGULARITIES

This chapter provides the beginning of the classification of the germs of holomorphic functions in three variables up to holomorphic equivalence: Two germs $f,g \in {}_n\mathcal{O}$ are called *holomorphically equivalent* $(f \sim g)$ if there is a biholomorphic germ $H: (\mathbb{C}^n,0) \to (\mathbb{C}^n,0)$ with $g \sim f \circ H$. Since equivalent germs have the same value at 0, attention is mostly restricted to germs f in the maximal ideal $\mathfrak{m} \subset {}_n\mathcal{O}$. The regular germs in $\mathfrak{m}$ form one equivalence class $\mathfrak{m} \setminus \mathfrak{m}^2$. For $f \in \mathfrak{m}^2$ we have $\operatorname{grad} f(0) = 0$. The germ f is said to have an isolated critical point if $\operatorname{grad} f$ does not vanish in a punctured neighbourhood $U \setminus \{0\}$. Germs with isolated critical points will only be considered for classification.

Our procedure requires a topology on $\mathcal{O}_n = \mathcal{O}$. It is defined in such a way that f is close to $0 \in \mathcal{O}$ if the Taylor coefficients of f are small. With respect to this topology the equivalence class $A_o = \mathfrak{m} \setminus \mathfrak{m}^2$ of regular germs is open and dense in $\mathfrak{m}$. The next complicated equivalence class A_1 is open and dense in $\mathfrak{m}^2$. It consists of all germs $f \in \mathfrak{m}^2$ such that the Hessian $(\partial^2 f/\partial z_i \partial z_j)$ has maximal rank n at 0. The equivalence classes of more complicated germs become thinner and thinner, even infinitely thin. A germ $f \in \mathfrak{m}$ with an isolated critical point is called *simple* if its equivalence class is not infinitely thin in the following sense: Finitely many equivalence classes form a neighbourhood of f in $\mathfrak{m}$. The equivalence classes of simple germs belong to two infinite series A_k (k=0,1,..) and D_k (k=4,5,..), and to one finite series E_k (k=6,7,8). They form the following hierarchy:

$$A_0 \leftarrow A_1 \leftarrow A_2 \leftarrow A_3 \leftarrow A_4 \leftarrow A_5 \leftarrow A_6 \leftarrow A_7 \leftarrow A_8 \leftarrow \cdot\cdot A_{k-1} \leftarrow \cdots$$

$$D_4 \leftarrow D_5 \leftarrow D_6 \leftarrow D_7 \leftarrow D_8 \leftarrow \cdots \leftarrow D_k \leftarrow \cdots$$

$$E_6 \leftarrow E_7 \leftarrow E_8$$

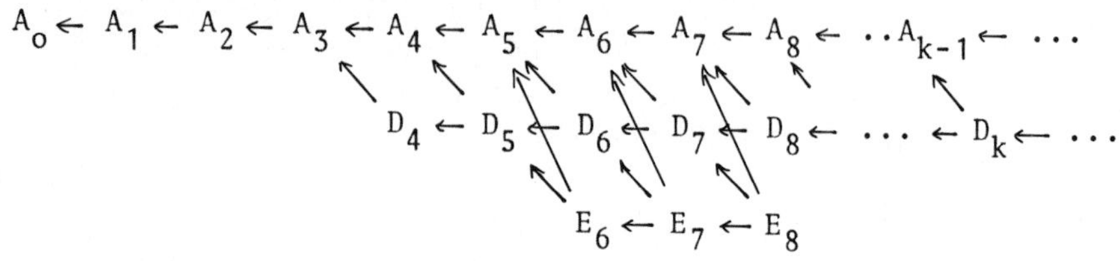

By following the arrows the germs become less and less complicated, more precisely: If a finite chain of arrows starts at
the class X and ends at the class Y (e.g. $X = E_7 \rightarrow E_6 \rightarrow D_5 \rightarrow A_4$
$= Y$), then X is contained in the closure of Y : For every
$f \in X$ there is a germ $g \in Y$ arbitrary close to f . This relation is called *adjacency* and is denoted by $X \succ Y$.

The handling of polynomials in two and three variables by
methods of high school mathematics (completing the square etc.)
will play an essential rôle in the establishment of the hierarchy of simple germs. But the efficacy of the elementary
methods relies on results of more advanced mathematics:

(1) All biholomorphic germs $H: (\mathbb{C}^n, 0) \rightarrow (\mathbb{C}^n, 0)$ form a group
$\mathcal{G}$. The holomorphic equivalence classes of germs $f \in {}_n\mathcal{O}$
are the orbits of the action of $\mathcal{G}$ on ${}_n\mathcal{O}$. This action is
studied by means of the finite dimensional reduction
$Q_k = \mathcal{O}/m^{k+1}$. The infinite dimensional group $\mathcal{G}$ acts on Q_k
via a finite dimensional linear algebraic group $\mathcal{G}_k < GL(Q_k)$.
General results about the *orbits of linear algebraic groups*
will be employed.

(2) The well known *multiplicity* of a zero of a holomorphic
function in one variable is generalized to the case of holo-
morphic germs $(\mathbb{C}^n, 0) \rightarrow (\mathbb{C}^n, 0)$. What interests here is the
multiplicity of the gradient map grad f of the germ $f \in {}_n\mathcal{O}$.

The combination of both methods reduces the investigation of
arbitrary germs f to their "leading terms" f_0 , which are
weighted homogeneous polynomials. They can finally be handled
(for two and three variables) by elementary methods.

The hierarchy of simple germs does not depend on the number n of variables provided $n \geq 2$. (For $n=1$ there is the A-series only.) The case $n=3$ is especially interesting:

A germ $f \in {}_3\mathfrak{m}$ is simple if and only if its zero set $N(f)$ is isomorphic to the germ of the quotient variety $\mathbb{C}^2/G$ of a finite subgroup $G < SL(2,\mathbb{C})$. The two infinite series A_k and D_k correspond to the cyclic and binary dihedral groups respectively, and E_6, E_7, E_8 correspond to the binary tetrahedral, octahedral, and icosahedral groups.

The hierarchy of simple singularities (germs) is due to V.I. Arnold (1972). Later he refined the methods of his first paper and extended the classification far beyond the simple germs. As reference the book by Arnold, Gusein-Zade, and Varchenko is recommended.

§1 Basic Concepts

The infinite dimensional $\mathbb{C}$-algebra $\mathcal{O} = {}_n\mathcal{O}_o$ is studied by means of the canonical projections

$$\pi_k : \mathcal{O} \to \mathcal{O}/\mathfrak{m}^{k+1} = Q_k \quad , \quad k=0,1,2,\ldots ,$$

onto the finite dimensional algebras Q_k. As any finite dimensional $\mathbb{C}$-vector space Q_k has a well defined topology. On $\mathcal{O}$ we used the *coarsest topology* such that all π_k are continuous: A basis for the open sets of $\mathcal{O}$ consists of all $\pi_k^{-1}(U)$, $k \in \mathbb{N}$ and $U \subset Q_k$ open. Thus $\mathcal{O}$ becomes a topological $\mathbb{C}$-algebra which satisfies the Hausdorff separation axiom. The maximal ideal $\mathfrak{m} \subset \mathcal{O}$ is closed.

The germs of holomorphic maps $H: (\mathbb{C}^n,0) \to (\mathbb{C}^n,0)$ form a $\mathbb{C}$-algebra $\mathcal{H}$. Since H is uniquely determined by its n components $h_1,\ldots,h_n \in \mathfrak{m}$, we identify $\mathcal{H} = \mathfrak{m}\times\ldots\times\mathfrak{m}$ (n times) and endow it with the product topology. Then $\mathcal{H}$ is a topological algebra. It acts continuously on $\mathcal{O}$ from right by composition,

$$\mathcal{O}\times\mathcal{H} \to \mathcal{O} \quad , \quad (f,H) \mapsto f\circ H \ ,$$

such that $\mathfrak{m}\mathcal{H} = \mathfrak{m}$. An element $H \in \mathcal{H}$ is said to have order k if $\mathfrak{m}H \subset \mathfrak{m}^k$. The maps of order k form a two sided ideal $\mathcal{H}^k \subset \mathcal{H}$. The quotient algebra

$$\mathcal{H}_k = \mathcal{H}/\mathcal{H}^{k+1} = Q_k\times\ldots\times Q_k$$

is finite dimensional. The projection is denoted by $\pi_k: \mathcal{H} \to \mathcal{H}_k$. The action of $\mathcal{H}$ on $\mathcal{O}$ induces an action of $\mathcal{H}_k$ on Q_k. We have $Q_1 = \mathcal{O}/\mathfrak{m} \oplus \mathfrak{m}/\mathfrak{m}^2 \cong \mathbb{C}^{n+1}$. The algebra $\mathcal{H}_1$ is the algebra $\mathrm{End}(\mathfrak{m}/\mathfrak{m}^2)$ of all linear endomorphisms of $\mathfrak{m}/\mathfrak{m}^2$. It acts as identity on the first factor $\mathcal{O}/\mathfrak{m}$ of Q_1. The determinant $\det: \mathrm{End}(\mathfrak{m}/\mathfrak{m}^2) \to \mathbb{C}$ yields the generalized determinant

$$\det{}_\infty: \mathcal{H} \xrightarrow{\pi_1} \mathcal{H}_1 = \mathrm{End}(\mathfrak{m}/\mathfrak{m}^2) \xrightarrow{\det} \mathbb{C} \ .$$

The Inverse Mapping Theorem says that $H \in \mathcal{H}$ is invertible if and only if $\det_\infty H \neq 0$. Thus the open subset

$$\mathcal{G} = \{H \in \mathcal{H} : \det H \neq 0\}$$

of $\mathcal{H}$ is the group of all germs of biholomorphic maps $(\mathbb{C}^n,0) \to (\mathbb{C}^n,0)$.

Two germs $f,g \in \mathcal{O}$ are said to be *holomorphically equivalent*, denoted by $f \sim g$, if and only if $g = f\circ H$ for some $H \in \mathcal{G}$. The investigation of all germs $f \in \mathcal{O}$ up to holomorphic equivalence is the investigation of the decomposition of $\mathcal{O}$ into orbits of the $\mathcal{G}$-action. Since the action leaves the value $f(0)$ invariant, attention is mostly restricted to germs $f \in \mathfrak{m}$.

The *hierarchy* of holomorphic germs is based on the following relation between germs $f,g \in \mathcal{O}$:
g is *adjacent* to f, denoted by $g \succ f$ or $f \prec g$, if and

only if g belongs to the closure of the $\mathcal{G}$-orbit through f, for short,

$$g \succ f \text{ if and only if } g \in \overline{f\mathcal{G}} \,.$$

Actually this is a relation between the orbits, viz. $g\mathcal{G} \subset \overline{f\mathcal{G}}$. Adjacency is reflexive $(f \succ f)$ and transitive ($f \succ g$ and $g \succ h$ imply $f \succ h$). Behind the relation $g \succ f$ there is the idea that g is more complicated than f.

A germ $f \in \mathfrak{m}$ is said to be *simple* if there is a neighbourhood U of f in $\mathfrak{m}$ such that finitely many $\mathcal{G}$-orbits only pass through U, in other word: The union of finitely many $\mathcal{G}$-orbits is a neighbourhood of f. Every $g \in U$ is also simple. If $g \succ f$ and g is simple, then f is simple.

§2 The Milnor Number

For a germ $f \in {}_n\mathcal{O}$ let $\Delta(f)$ denote the ideal generated by the germs of the partial derivatives $\partial f/\partial z_1, \ldots, \partial f/\partial z_n$,

$$\Delta(f) = \langle \partial f/\partial z_1, \ldots, \partial f/\partial z_n \rangle \,.$$

The *Milnor number* $0 \le \mu(f) \le \infty$ is defined to be the dimension of the complex quotient vector space $\mathcal{O}/\Delta(f)$,

$$\mu(f) = \dim_{\mathbb{C}} \mathcal{O}/\Delta(f) \,.$$

Remark: The name "Milnor number" comes from Milnor's topological characterization of $\mu=\mu(f)$: Let f be defined in a neighbourhood of the closed ε-ball B_ε around 0 in $\mathbb{C}^n$. If f is regular (grad $f \neq 0$) at all points $z \in B_\varepsilon \setminus \{0\}$, then for every c close to but different from $f(0)$ the fibre $\{z \in B_\varepsilon : f(z)=c\}$ is a compact bounded $(2n-2)$-dimensional manifold which has the homotopy type of a bouquet $S^{n-1} \vee \ldots \vee S^{n-1}$ (one-point union) of μ $(n-1)$-spheres, see Milnor (1).

In this section first an arbitrary ideal $\mathfrak{a} \subset \mathfrak{m}$ is considered.

Later we specialize to $\mathfrak{a} = \Delta(f)$. The following statements are equivalent:

(1) $\dim \mathcal{O}/\mathfrak{a} < \infty$

(2) $\mathfrak{m}^k \subset \mathfrak{a}$ for some k

(3) $N(\mathfrak{a}) = \{0\}$.

This has been proved in III§8, proposition 2. A refined version of the argument used there yields

Proposition 1: *If* $\dim \mathcal{O}/(\mathfrak{a}+\mathfrak{m}^{k+1}) \leq k$, *then* $\mathfrak{m}^k \subset \mathfrak{a}$.

Proof: There are k epimorphisms

$$\mathcal{O}/(\mathfrak{a}+\mathfrak{m}^{k+1}) \to \mathcal{O}/(\mathfrak{a}+\mathfrak{m}^{k}) \to \dots \to \mathcal{O}/(\mathfrak{a}+\mathfrak{m}) = \mathcal{O}/\mathfrak{m}$$

which lower the dimension from $\dim \mathcal{O}/(\mathfrak{a}+\mathfrak{m}^{k+1}) \leq k$ to $\dim \mathcal{O}/\mathfrak{m} = 1$. Therefore at least one epimorphism is an isomorphism, i.e. $\mathfrak{m}^j \subset \mathfrak{a}+\mathfrak{m}^{j+1}$ for some $1 \leq j \leq k$. Then by Nakayama's lemma, compare III§5, $\mathfrak{m}^k \subset \mathfrak{m}^j \subset \mathfrak{a}$.

The proposition is also used in the weaker form

(4) $$\mathfrak{m}^k \subset \mathfrak{a} \quad \text{for} \quad k \geq \dim \mathcal{O}/\mathfrak{a} .$$

In the following sections the quotient $\mathcal{O}/\mathfrak{m}\mathfrak{a}$ will occur. It is related to $\mathcal{O}/\mathfrak{a}$ by the exact sequence

(5) $$0 \to \mathfrak{a}/\mathfrak{m}\mathfrak{a} \to \mathcal{O}/\mathfrak{m}\mathfrak{a} \to \mathcal{O}/\mathfrak{a} \to 0$$

and by the following

Lemma 2: *The germs* $f_1,\dots f_r$ *generate the ideal* $\mathfrak{a}$ *if and only if their residue classes span the vector space* $\mathfrak{a}/\mathfrak{m}\mathfrak{a}$. In particular, $\gamma(\mathfrak{a}) = \dim \mathfrak{a}/\mathfrak{m}\mathfrak{a}$ is the minimal number of generators of $\mathfrak{a}$ and

(6) $$\dim \mathcal{O}/\mathfrak{m}\mathfrak{a} = \gamma(\mathfrak{a}) + \dim \mathcal{O}/\mathfrak{a} .$$

Proof: If $\mathfrak{a} = \langle f_1,\dots,f_r \rangle$, every $f \in \mathfrak{a}$ has the form $f = \Sigma h_i f_i$ with $h_i \in \mathcal{O}$. Let $\alpha_i = h_i(0) \in \mathbb{C}$. Then $f = \Sigma \alpha_i f_i + \Sigma(h_i - \alpha_i) f_i$ with $h_i - \alpha_i \in \mathfrak{m}$ and so $f = \Sigma \alpha_i f_i$ modulo $\mathfrak{m}\mathfrak{a}$. Vice versa let $\mathfrak{b} = \langle f_1,\dots,f_r \rangle$ and let the residue classes of $f_1,\dots,f_r$ span $\mathfrak{a}/\mathfrak{m}\mathfrak{a}$. Then $\mathfrak{a} \subset \mathfrak{b} + \mathfrak{m}\mathfrak{a}$ and so by Nakayama's lemma, $\mathfrak{a} \subset \mathfrak{b}$. The formula (6) follows from (5) . It includes the possibility that both dimensions are infinite.

A germ $f \in {}_n\mathcal{O}$ is called *regular* if grad $f(0) \neq 0$. Any two regular germs are almost holomorphically equivalent, to wit: *If both* f *and* g *are regular, then* $f-f(0)$ *and* $g-g(0)$ *are holomorphically equivalent.* This is due to the fact that each regular $f \in \mathfrak{m}$ is the first component of a biholomorphic germ $F: (\mathbb{C}^n,0) \to (\mathbb{C}^n,0)$ and so is holomorphically equivalent to the first coordinate z_1.

A germ $f \in {}_n\mathcal{O}$ is called *singular* if it is not regular, equivalently grad $f(0)=0$ or $\Delta(f) \subset \mathfrak{m}$. We are particularly interested in germs f with an isolated singularity. This means grad $f(0)=0$, but grad f nowhere vanishes on a punctured neighbourhood $U \setminus \{0\}$. This is equivalent to $N(\Delta(f)) = \{0\}$ and hence to $\mu(f) = \dim \mathcal{O}/\Delta(f) < \infty$, compare (1) and (3) above.

The minimal number of generators $\gamma(\Delta(f)) \leq n$. If f has an isolated singularity, $\gamma(\Delta(f))=n$ because $0 = \dim N(\Delta(f) \geq n-\gamma(\Delta(f))$, compare III§9. Therefore (6) yields

$$\dim {}_n\mathcal{O}/\mathfrak{m}\Delta(f) = n+\mu(f) \quad , \tag{8}$$

including the case that both sides are infinite.

<u>Proposition 3</u>: *The Milnor number is semi-continuous with respect to the topology of* $\mathcal{O}$,

$$\mu(g) \leq \mu(f) \quad \textit{for all } g \textit{ close to } f .$$

This is a consequence of the following

<u>Lemma 4</u>: *Let the residue classes of* $h_1,\ldots,h_r \in \mathcal{O}$ *span the vector space* $\mathcal{O}/\Delta(f)$. *Then they also span* $\mathcal{O}/\Delta(g)$ *for all* g *close to* f.

Proof: Let $k \geq r+1$. The subspace $(\Delta(f)+\mathfrak{m}^{k+1})/\mathfrak{m}^{k+1}$ of $Q_k = \mathcal{O}/\mathfrak{m}^{k+1}$ is spanned by the residue classes of $z^\alpha \partial f/\partial z_i$ for all $|\alpha| \leq k$ and $1 \leq i \leq n$. Together with $h_1,\ldots,h_r$ they span Q_k. The maps $g \mapsto z^a \partial g/\partial z_i$ modulo $\mathfrak{m}^{k+1}$ are continuous. Therefore the residue classes modulo $\mathfrak{m}^{k+1}$ of all $z^\alpha \partial g/\partial z_i$ and of $h_1,\ldots,h_r$ also span Q_k provided g is close to f. The classes $z^\alpha \partial g/\partial z_i$ span $(\Delta(g)+\mathfrak{m}^{k+1})/\mathfrak{m}^{k+1}$.

The exact sequence $0 \to (\Delta(g)+\mathfrak{m}^{k+1})/\mathfrak{m}^{k+1} \to Q_k \to$

$\to \mathcal{O}/(\Delta(g)+\mathfrak{m}^{k+1}) \to 0$ implies that the classes of $h_1,\ldots,h_r$ span $\mathcal{O}/(\Delta(g)+\mathfrak{m}^{k+1})$. In particular, $\dim \mathcal{O}/(\Delta(g)+\mathfrak{m}^{k+1}) \leq r \leq k+1$. Therefore by proposition 1 $\mathcal{O}/\Delta(g) = \mathcal{O}/(\Delta(g)+\mathfrak{m}^{k+1})$.

Proposition 5: *If the germs* f *and* g *are holomorphically equivalent, they have the same Milnor number.*

Proof: Let $g = f\circ H$ for some biholomorphic $H: (\mathbb{C}^n,0) \to (\mathbb{C}^n,0)$. Then H induces an automorphism $H^o: {}_n\mathcal{O} \to {}_n\mathcal{O}$ with $\Delta(g) \subset H^o(\Delta(f))$ because of $\partial g/\partial z_j = \sum_i (\partial f/\partial z_i \circ H)\cdot \partial h_i/\partial z_j$. Let G be the inverse of H. Then G^o is the inverse of H^o and $\Delta(f) \subset G^o(\Delta(g))$. Hence $H^o(\Delta(f)) \subset \Delta(g)$ and so $H^o(\Delta(f)) = \Delta(g)$.

§3 Transformation Groups

The *Zariski topology* on a finite dimensional $\mathbb{C}$-vector space V is defined by means of the following closed sets: A subset $A \subset V$ is *Zariski closed* if it is the common set of zeros of finitely many polynomials. The Zariski topology is coarser than the usual complex topology. The Hausdorff separation axiom does not hold.

The finite dimensional $\mathbb{C}$-algebras $\mathcal{H}_k$, which have been introduced in §1, are related by canonical projections

$$\pi_{1,k}: \mathcal{H}_k \to \mathcal{H}_1 \quad \text{if } 1 \leq k .$$

The generalized determinant

$$\det_k: \mathcal{H}_k \xrightarrow{\pi_{1,k}} \mathcal{H}_1 = \mathrm{End}(\mathfrak{m}/\mathfrak{m}^2) \xrightarrow{\det} \mathbb{C}$$

is a polynomial function. Therefore the group

$$\mathcal{G}_k = \{\eta \in \mathcal{H}_k : \det_k \eta \neq 0\}$$

of all invertible elements of $\mathcal{H}_k$ is Zariski open in $\mathcal{H}_k$. The product and the inverse are algebraic maps, thus $\mathcal{G}_k$ is an algebraic group. It acts algebraically on Q_k. The following general result will be used:

Proposition 1: *Each orbit of an algebraic group action is a smooth locally closed subset, whose boundary is a union of orbits of strictly lower dimension.*

Remarks concerning the proposition: A subset of a topological space is said to be locally closed if it is the intersection of an open set with a closed set. The boundary of a subset A is 'closure A minus A'. In the proposition everything refers to the Zariski topology. For more informations about algebraic group actions and for the proof of the proposition the reader may consult J.E. Humphrey's book, in particular p.60.

A complex algebraic group is also a complex Lie group, in our case: The group $\mathcal{G}_k$ is a finite dimensional complex Lie group with Lie algebra $\mathcal{H}_k$. The Lie bracket is the commutator $[\alpha,\beta] = \alpha\circ\beta-\beta\circ\alpha$; but we do not need this. Let $\varphi \in Q_k$ be fixed. The orbit map

$$(1) \qquad \mathcal{G}_k \to Q_k \ , \ \gamma \to \varphi\circ\gamma \ ,$$

is holomorphic. Its image is the orbit $\varphi\mathcal{G}_k$ through φ. The differential (at $id \in \mathcal{G}_k$, where $\mathcal{H}_k$ is the tangent space) is the linear map

$$(2) \qquad \mathcal{H}_k \to Q_k \ , \ \eta \to \lim_{\lambda\to 0} (\varphi\circ(id+\lambda\eta)-\varphi)/\lambda \quad , \quad \lambda \in \mathbb{C} \ .$$

The image of this differential is the tangent space $T_\varphi(\varphi\mathcal{G}_k)$ of the orbit $\varphi\mathcal{G}_k$ at φ.

The finite dimensional orbit map (1) and its differential (2) are induced by the *infinite dimensional orbit map*

$$(3) \qquad \mathcal{G} \to \mathcal{O} \ , \ G \to f\circ G \ , \ \text{where } \varphi = \pi_k(f) \ ,$$

and its differential

$$(4) \qquad \mathcal{H} \to \mathcal{O} \ , \ H \to \lim_{\lambda\to 0} (f\circ(id+\lambda H)-f)/\lambda = \sum_{i=1}^{n} h_i \cdot \partial f/\partial z_i \quad ,$$

where h_i is the i-th component of H. The image of this differential

(5) $$T_f(f\mathcal{G}) = \mathfrak{m}\Delta(f)$$

is called the *tangent space* of $f\mathcal{G}$ at f. Since $T_\varphi(\varphi\mathcal{G}_k) = \pi_k(T_f(f\mathcal{G}))$ we obtain

(6) $$T_\varphi(\varphi\mathcal{G}_k) = (\mathfrak{m}\Delta(f)+\mathfrak{m}^{k+1})/\mathfrak{m}^{k+1} .$$

Every $\gamma\in\mathcal{G}_k$ determines an automorphism of Q_k which maps $T_\varphi(\varphi\mathcal{G}_k)$ isomorphically onto $T_{\varphi\gamma}(\varphi\mathcal{G}_k)$. In particular

(7) $$\dim(\varphi\mathcal{G}_k) = \dim T_\psi(\psi\mathcal{G}_k)$$

does not depend on the choice of $\psi\in\varphi\mathcal{G}_k$. The quotient spaces

$$N_f(f\mathcal{G}) := \mathcal{O}/T_f(f\mathcal{G}) = \mathcal{O}/\mathfrak{m}\Delta(f) ,$$

$$N_\varphi(\varphi\mathcal{G}_k) := Q_k/T_\varphi(\varphi\mathcal{G}_k) = \mathcal{O}/(\mathfrak{m}\Delta(f)+\mathfrak{m}^{k+1}) \quad , \varphi=\pi_k(f)$$

are called the *normal spaces* of the orbits at f and φ resp. In contrast to the tangent space $N_f(f\mathcal{G})$ has a chance to be finite dimensional: By §2(8)

(8) $$\operatorname{codim} f = \dim N_f(f\mathcal{G}) = n + \mu(f) .$$

This is finite if and only if f has an isolated singularity. If $u_1,\dots,u_\mu$ represent a basis of $\mathcal{O}/\Delta(f)$ then

(9) $$u_1,\dots,u_\mu , \partial f/\partial z_1,\dots, \partial f/\partial z_n$$

represent a basis of $N_f(f\mathcal{G})$. We define

(10) $$\operatorname{codim}_k f = \dim N_\varphi(\varphi\mathcal{G}_k) = \dim \mathcal{O}/(\mathfrak{m}\Delta(f)+\mathfrak{m}^{k+1}) .$$

Then $\operatorname{codim}_k f \leq \operatorname{codim} f$, and $\operatorname{codim}_k f = \operatorname{codim} f < \infty$ if and only if $\mathfrak{m}^{k+1}\subset\mathfrak{m}\Delta(f)$.

§4 Families of Germs

Let $T \subset \mathbb{C}^r$ be an open and connected subset. The coordinates are denoted by $t=(t_1,\ldots,t_r)$. A function

$$F: T \to {}_n\mathcal{O} \quad , \quad t \mapsto f_t \ ,$$

is called a *holomorphic family* of germs with r parameters if there is a holomorphic function, also denoted by F, defined in a neighbourhood of $\{0\}\times T \subset \mathbb{C}^n\times\mathbb{C}^r$ such that

$$f_t = \text{germ of } (z \mapsto F(z,t)) \ .$$

The family is called *algebraic* if in the Taylor expansion $f_t = \sum_\alpha a_\alpha(t)z^\alpha$ the coefficients $a_\alpha(t)$ are polynomials in $t \in \mathbb{C}^r = T$.

Let $F_1,\ldots,F_q : T \to \mathcal{O}$ be q holomorphic [algebraic] families. For every $t \in T$ they determine q germs $f_{1,t},\ldots, f_{q,t} \in \mathcal{O}$. Let $\mathfrak{a}_t = \langle f_{1,t},\ldots,f_{q,t}\rangle$ denote the ideal generated by them. Let

$$m_t = \dim \mathcal{O}/\mathfrak{a}_t \quad , \quad T_k = \{t \in T : m_t \geq k\}$$

Then $T = T_0 \supset T_1 \supset \ldots \supset T_k \supset \ldots \supset T_\infty$.

Proposition 1: *Every* T_k *is an analytic [Zariski closed] subset of* T .

For the proof the following consideration is used: Let V denote an n-dimensional complex vector space, $n < \infty$. For any $(v_1,\ldots,v_s) \in V\times\ldots\times V$ (s times) let $\operatorname{span}(v_1,\ldots,v_s) \subset V$ denote the subvector space spanned by $v_1,\ldots,v_s$. If a base of V is fixed, $(v_1,\ldots,v_s)$ is a $(s\times n)$-matrix, and $\dim \operatorname{span}(v_1,\ldots,v_s)$ is the rank of this matrix. Therefore $\dim \operatorname{span}(v_1,\ldots,v_s) \leq d$ if and only if all $(d\times d)$-subdeterminants of the matrix vanish. These are polynomial conditions. Therefore the set

$$\{(v_1,\ldots,v_s) \in V\times\ldots\times V : \dim \operatorname{span}(v_1,\ldots,v_s) \leq d\}$$

is Zariski closed in $V\times\ldots\times V$. We shall use this result in the following version

Lemma 2: *The subset*

$$\{(v_1,..,v_s) \in V\times..\times V:\quad \dim V/\mathrm{span}(v_1,..,v_s) \geq k\}$$

is Zariski closed in $V\times..\times V$ *for every* $k=0,1,\dots$.

Proof of proposition 1: For every multi-index α with $|\alpha| < k$ and every $i=1,\dots,q$ the holomorphic [polynomial] map $T \to Q_k = \mathcal{O}/\mathfrak{m}^{k+1}$, $t \mapsto z^\alpha f_{i,t} \bmod \mathfrak{m}^{k+1}$, is considered. The number s of these maps is finite. Let $L_t = \mathrm{span}\{z^\alpha f_{i,t} \bmod \mathfrak{m}^{k+1}\} \subset Q_k$. By lemma 2, the subset

$$T_k^* = \{t \in T:\quad \dim Q_k/L_t \geq k\}$$

is analytic [Zariski closed]. Now $Q_k/L_t = \mathcal{O}/(\mathfrak{a}_t+\mathfrak{m}^{k+1})$. By §2, proposition 1, the conditions $\dim \mathcal{O}/\mathfrak{a}_t \geq k$ and $\dim \mathcal{O}/(\mathfrak{a}_t+\mathfrak{m}^{k+1}) \geq k$ are equivalent. Therefore $T_k^* = T_k$.

The *differential* at t of the holomorphic family $F: T \to \mathcal{O}$ is defined to be the linear map

(1) $$D_tF: \mathbb{C}^r \to \mathcal{O} \quad , \quad v \to \sum_{j=1}^{r} v_j \cdot \partial f_t/\partial t_j \quad ,$$

where $\partial f_t/\partial t_j$ denotes the germ of $z \mapsto (\partial F/\partial t_j)(z,t)$ for fixed t . The maps .

(2) $$F_k = \pi_k \circ F: T \to \mathcal{O} \to \mathcal{O}/\mathfrak{m}^{k+1} = Q_k \quad , \quad k=0,1,\dots$$

are holomorphic. The differential of F_k is induced by (1) , viz.

(3) $$D_tF_k = \pi_k \circ D_tF \quad .$$

Proposition 3: *Assume that all* f_t *are pairwise holomorphically equivalent. Then*

(4) $$\partial f_t/\partial t_j \in \mathfrak{m}\Delta(f_t) \quad \textit{for every} \quad t \in T \quad \textit{and} \quad j=1,\dots,r \ .$$

In other words: If $F(T)$ is contained in one $\mathcal{G}$-orbit, then $D_tF(\mathbb{C}^r)$ is contained in $T_{f_t}(f_t\mathcal{G})$. In this form the statement is obviously true in the finite dimensional cases "modulo $\mathfrak{m}^{k+1}$" , therefore $\partial f_t/\partial t_j \in \mathfrak{m}\Delta(f_t)+\mathfrak{m}^{k+1}$ for every k . The intersection over all k yields (4) .

Proposition 3 has a partial converse. In order to formulate it, the following notations are introduced: Fix an $s \in T$ and denote the local algebra of holomorphic functions at $(0,s)\in\mathbb{C}^n\times T$ by $\mathcal{O}'$. Let $\Delta(F)\subset\mathcal{O}'$ denote the ideal generated by $\partial F/\partial z_i$, $i=1,\dots,n$. Consider $\mathcal{O}$ as subalgebra of $\mathcal{O}'$ and consider the ideal $\mathfrak{m}\Delta(F)\subset\mathcal{O}'$ where $\mathfrak{m}\subset\mathcal{O}$ denotes the maximal ideal.

<u>Proposition 4</u>: *Let* $F: T\to\mathcal{O}$ *be a holomorphic one-parameter family. If for every fixed* $s\in T$ *the germ* $\partial F/\partial t\in\mathcal{O}'$ *is contained in the ideal* $\mathfrak{m}\Delta(F)$, *then all* f_t *for* $t\in T$ *are mutually holomorphically equivalent.*

Proof: Since T is connected, it suffices to find a $\mathbb{C}^n$-valued holomorphic map H, defined in a neighbourhood of $(0,s)\in\mathbb{C}^n\times T$, such that

(5) $\quad F(z,t)=F(H(z,t),s)\ ,\quad H(0,t)=0\ ,\quad H(z,s)=z\ .$

Then for all t, close to s, the map $z\mapsto H(z,t)$ is a biholomorphic germ $H_t:(\mathbb{C}^n,0)\to(\mathbb{C}^n,0)$ with $f_t=f_s\circ H_t$.

By assumption $\partial F/\partial t=\Sigma\, g_i\partial F/\partial z_i$ where $g_i\in\mathfrak{m}\mathcal{O}'$, i.e. g_i is a holomorphic function defined in a neighbourhood of $(0,s)\in\mathbb{C}^n\times T$ and $g_i(0,t)=0$. We obtain $h_i(z,t)$ so that $\partial h_i/\partial t=g_i$ and $h_i(z,s)=z$ by integration with respect to t. Then $h_i(0,t)=0$ because $g_i(0,t)=0$ and $h_i(0,s)=0$. Let H be the map with components h_i. It fullfils (5): The second and third equation have just been obtained. The first equation is true for $t=s$. Since both sides have the same derivative with respect to t, the equation is true for all t close to s.

§5 Finitely Determined Germs

The germ $f\in\mathcal{O}$ is said to be k-*determined* if every $g\in\mathcal{O}$ with $\pi_k(g)=\pi_k(f)\in\mathcal{O}/\mathfrak{m}^{k+1}=Q_k$ is holomorphically equivalent to f. Actually this is a property of the orbit $\pi_k(f)\mathcal{G}_k$:

f *is* k-*determined if and only if* $\pi_k^{-1}(\pi_k(f)\mathcal{G}_k)=f$.

Let $h \in \mathfrak{m}^{k+1}$. If f is k-determined, all members of the one-parameter family $f_t = f + th$ are pairwise holomorphically equivalent and so by §4, proposition 3, $h \in \mathfrak{m}\Delta(f)$. This proves

<u>Proposition 1</u>: *If* f *is* k-*determined, then* $\mathfrak{m}^{k+1} \subset \mathfrak{m}\Delta(f)$, *hence* $\mathrm{codim}_k f = \mathrm{codim}\, f$.

The converse is almost true,

<u>Theorem 2</u>: *If* $\mathfrak{m}^{k+1} \subset \mathfrak{m}^2\Delta(f)$, *then* f *is* k-*determined. In particular, every germ with an isolated singularity is finitely determined.*

Proof: Let $h \in \mathfrak{m}^{k+1}$ be given. If suffices to show that all $f_t = f + th$ for $t \in \mathbb{C}$ are pairwise holomorphically equivalent. By §4, proposition 4, this is the case as soon as $h \in \mathfrak{m}\Delta(F) \subset \mathcal{O}' = \mathcal{O}_{(0,s)}$ for every s . Now $h \in \mathfrak{m}^{k+1}$ implies $\partial h/\partial z_i \in \mathfrak{m}^k$ and so $t \cdot \partial h/\partial z_i \in \mathfrak{m}^k \mathcal{O}' \subset \mathfrak{m}'^{k-1}\mathcal{O}'$, where $\mathfrak{m}' \subset \mathcal{O}'$ denotes the maximal ideal. Since $\partial F/\partial z_i = \partial f/\partial z_i + t\partial h/\partial z_i$, we obtain $\Delta(f)\,\mathcal{O}' \subset \Delta(F) + \mathfrak{m}'\mathfrak{m}^{k-1}\mathcal{O}'$. Thus the assumption $\mathfrak{m}^{k+1} \subset \mathfrak{m}^2\Delta(f)$ yields $\mathfrak{m}^{k+1}\mathcal{O}' \subset \mathfrak{m}^2\Delta(F) + \mathfrak{m}'\mathfrak{m}^{k+1}\mathcal{O}'$ and so by Nakayama's lemma, $\mathfrak{m}^{k+1}\mathcal{O}' \subset \mathfrak{m}^2\Delta(F)$. This implies $h \in \mathfrak{m}^{k+1} \subset \mathfrak{m}^{k+1}\mathcal{O}' \subset \mathfrak{m}^2\Delta(F) \subset \mathfrak{m}\Delta(F)$.

<u>Corollary 3</u>: *If* $k \geq \mu(f)+1$, *then* f *is* k-*determined.*

Proof: The assumption implies $\mathfrak{m}^{k-1} \subset \Delta(f)$, compare §2(4), and so $\mathfrak{m}^{k+1} \subset \mathfrak{m}^2\Delta(f)$.

<u>Proposition 4</u>: *Let* $f,g \in \mathfrak{m}$. *If* g *is adjacent to* f $(f \prec g)$, *then* $\mu(f) \leq \mu(g)$. *If* $f \prec g$ *and* $\mu(f) = \mu(g) < \infty$, *then* f *and* g *are holomorphically equivalent* $(f \sim g)$.

Proof: The first statement follows from the semi-continuity of μ , see §2, proposition 3. Fix some $k \geq \mu(f)+1 = \mu(g)+1$. Then $\mathfrak{m}^{k-1} \subset \Delta(f) \cap \Delta(g)$ by §2(4) and so $\mathrm{codim}_k = \mathrm{codim}$ for **f** and g by the last statement of §3. The orbits $\varphi\mathcal{G}_k$ through $\varphi = \pi_k(f)$ and $\psi\mathcal{G}_k$ through $\psi = \pi_k(g)$ have the same dimension. Since $f \prec g$, the orbit $g\mathcal{G}$ is contained in the closure of $f\mathcal{G}$, and thus $\psi\mathcal{G}_k$ is contained in the closure of $\varphi\mathcal{G}_k$. Here "closure" refers to the usual topology. Since the Zariski topology is coarser, $\psi\mathcal{G}_k$ is also contained in the Zariski closure of $\varphi\mathcal{G}_k$. Then §3, proposition 1, implies $\psi\mathcal{G}_k = \varphi\mathcal{G}_k$. By Corollary 3, $\pi_k^{-1}(\varphi\mathcal{G}_k) = f\mathcal{G}$, and so $g \in f\mathcal{G}$.

Proposition 5: *Let* $f \in \mathfrak{m}$ *be simple and* $\mu(f) < \infty$. *There is a neighbourhood* U *of* f *in* $\mathfrak{m}$ *such that* $g \in U$ *and* $\mu(g) = \mu(f)$ *imply* $g \sim f$.

Proof: Since f is simple, there are finitely many $\mathcal{G}$-orbits $L_1,..,L$ such that $U=L_1 \cup ... \cup L_r$ is a neighbourhood of f . If $f \in \overline{L_r}$ (closure of L_r) , the remaining union $L_1 \cup ... \cup L_{r-1}$ is still a neighbourhood of f . Thus we may assume that $f \in \overline{L_i}$ for each $i=1,...,r$. Let $g \in U$. Then $g \in L_i$ for some i and so $g \prec f$. This implies $g \sim f$ by the preceding proposition.

§6 Unfoldings

A holomorphic r-parameter family $F: T \to \mathcal{O}$, $t \mapsto f_t$, is called an *unfolding* of $f \in \mathcal{O}$ if $f=f_s$ for some $s \in T$. For simplicity $s=0$ is assumed. By means of the differential D_oF , compare §4(1), the *normal map* of the unfolding is defined to be the linear map

$$(1) \qquad N_oF: \mathbb{C}^r \xrightarrow{D_oF} \mathcal{O} \to \mathcal{O}/\mathfrak{m}\Delta(f) = N_f(f\mathcal{G}) .$$

The unfolding is called *miniverse (transverse, universe)* if N_oF is injective (surjective, bijective). A transverse unfolding requires $r \geq \mu(f)+n$, in particular $\mu(f) < \infty$.

Proposition 1: *If* $t \mapsto f_t$ *is a miniverse unfolding of* $f=f_0$ *and* $\mu(f) < \infty$, *then* $0 \in T$ *is an isolated point of* $\{t \in T : f_t \sim f\}$.

Proof: Let $k \geq \mu(f)$ be fixed. Then $\mathfrak{m}^k \subset \Delta(f)$ by §2(4) and so $\mathrm{codim}_k f = \mathrm{codim}\, f$ be the end of §3. Let $\varphi = \pi_k(f)$. Then $N_f(f\mathcal{G}) = N_\varphi(\varphi\,\mathcal{G}_k)$ and so the differential of F_k: $T \to \mathcal{O} \to Q_k$ satisfies $(D_o F_k)^{-1}(T_\varphi(\varphi\,\mathcal{G}_k)) = \{0\}$. Therefore by the Inverse Mapping Theorem, $0 \in T$ is an isolated point of $F_k^{-1}(\varphi\,\mathcal{G}_k) \supset F^{-1}(f\mathcal{G}) = \{t \in T\colon f_t \sim f\}$.

Example: Let $u \in \mathcal{O}$ represent a non-zero element of $\mathcal{O}/\mathfrak{m}\Delta(f)$. Then $f+tu$ for $t \in \mathbb{C}$ is a one-parameter miniverse unfolding of $f=f_0$ and so $f+tu \not\sim f$ for all t close to 0 but $t \neq 0$.

Proposition 2: *If* $F\colon T \to \mathcal{O}$ *is a transverse unfolding of* f *, then* $F(T)\mathcal{G}$ *is a neighbourhood of* f *in* $\mathcal{O}$.

Proof: Consider the map

$$\alpha\colon T\times\mathcal{G} \to \mathcal{O} \quad , \quad \alpha(t,\Phi) = f_t \circ \Phi \quad ,$$

and the induced maps in finite dimensions

$$\alpha_k\colon T\times\mathcal{G}_k \to Q_k \ , \quad \alpha_k(t,\pi_k(\Phi)) = \pi_k(\alpha(t,\Phi)) \quad .$$

The differential of α is defined to be the linear map

$$D\alpha\colon \mathbb{C}^r \times \mathcal{H} \to \mathcal{O} \quad ,$$

$$(v_1,\ldots,v_r,h_1,\ldots h_n) \to \sum_{j=1}^{r} v_j \partial f_0/\partial t_j + \sum_{i=1}^{n} h_i \partial f/\partial z_i \ .$$

Here $\partial f_0/\partial t_j$ denotes the germ of $z \mapsto \partial F(z,0)/\partial t_j$, compare §4(1). Since F is transverse, $D\alpha$ is surjective. Therefore for all k the induced differentials at $(0,\mathrm{id})$

$$D\alpha_k \ : \mathbb{C}^r \times \mathcal{H}_k \to Q_k \quad , \ D\alpha_k(v,\pi_k(H)) = \pi_k \circ D\alpha(t,H)$$

are surjective, too. Then α_k is a submersion, compare III§3, and so α_k is open at $(0,\mathrm{id})$. If t is close to 0 , the Milnor number $\mu(f_t) \leq \mu(f) < \infty$. Thus there is a neighbourhood T_0 of 0 in T and there is a k , such that f_t is k-determined for $t \in T_0$. Since α_k is open at $(0,\mathrm{id})$, the set $F_k(T_0)\mathcal{G}_k = \alpha_k(T_0 \times \mathcal{G}_k)$ is a neighbourhood of $\pi_k(f)$ in Q_k, and so the inverse image $\pi_k^{-1}(F_k(T_0)\mathcal{G}_k)$ is a neighbourhood of f in $\mathcal{O}$. The proof will be finished as soon as the inclusion

$$\pi_k^{-1}(F_k(T_0)\mathcal{G}_k) \subset F(T_0)\mathcal{G}$$

is obtained. Let $g \in \pi_k^{-1}(F_k(T_0)\mathcal{G})$. Then $\pi_k(g)=\pi_k(f_t\Phi)$ for some $t \in T_0$ and some $\Phi \in \mathcal{G}$. Since f_t is k-determined, this implies $g \sim f_t\Phi \subset F(T_0)$.

Corollary 3: *Let* $u_1,\dots,u_\mu \in {}_n\mathcal{O}$ *represent a base of* $\mathcal{O}/\Delta(f)$, *where* $\mu < \infty$. *Consider*

$$g_t = f + t_1u_1 + \dots + t_\mu u_\mu + t_{\mu+1}\partial f/\partial z_1 + \dots + t_{\mu+n}\partial f/\partial z_n \quad ,$$

$$t=(t_1,\dots,t_{\mu+n}) \in \mathbb{C}^{\mu+n} .$$

Let V *be a neighbourhood of* 0 *in* $\mathbb{C}^{\mu+n}$. *Then* $\{g \in \mathcal{O}:$ $g \sim g_t$ for some $t \in V\}$ *is a neighbourhood of* f *in* $\mathcal{O}$. *In particular,* f *is simple if and only if the* g_t *in* $\mathfrak{m}$ *for* $t \in V$ *belong to finitely many orbits.*

Proof: The elements $u_1,\dots,u_\mu, \partial f/\partial z_1,\dots, \partial f/\partial z_n$ represent a base of $N_f(f\mathcal{G})$, compare §2, lemma 2. Therefore the unfolding $t \mapsto g_t$ of f is universe, and so the corollary follows from proposition 2.

Proposition 4: *Let* $F: T \to \mathfrak{m}$, $t \mapsto f_t$, *be an unfolding of the simple germ* f *such that* $\mu(f_t) = \mu(f) < \infty$ *is constant throughout* T. *Then* $f_t \sim f$ *for every* $t \in T$.

Proof: By §5, proposition 5, $F^{-1}(f\mathcal{G})$ is a neighbourhood of 0 in T. The claim is $F^{-1}(f\mathcal{G})=T$. Let $k \geq \mu(f)+1$ so that every f_t is k-determined. Let $\varphi = \pi_k(f) \in Q_k$ and let $F_k = \pi_k \circ F$. Then $F_k^{-1}(\varphi\mathcal{G}_k) = F^{-1}(f\mathcal{G})$. By §3, proposition 1 there are Zariski closed subsets A_k and B_k in Q_k such that $\varphi\mathcal{G}_k = A_k \setminus B_k$. Then $A = F_k^{-1}(A_k)$ and $B = F_k^{-1}(B_k)$ are analytic subsets of T and $F^{-1}(f\mathcal{G}) = A \setminus B$. This is a neighbourhood of 0, so $A = T$ and $F^{-1}(f\mathcal{G})$ is dense in T. Then $f_t \succ f$ for every $t \in T$, whence $f_t \sim f$ by §5, proposition 4.

§7 The Multiplicity

First some topological methods for the complex analysis of several variables are presented which are well known for one variable. For the requisites from differential topology, Hirsch's book, Chap.5, Sec.1, is recommended.

Let $M \subset \mathbb{C}^n$ be a compact oriented differentiable submanifold of real dimension $2n-1$ without boundary. Let $F: M \to \mathbb{C}^n$ be a continuous map with $F(z) \neq 0$ throughout M. Then $F/\|F\|: M \to S^{2n-1} = \{w: \|w\| = 1\}$ has a degree $\in \mathbb{Z}$. For simplicity this is called the *degree* of F. It has the following properties:

(1) If $M = M_1 \dot{\cup} M_2$ is a disjoint union, the degree of F on M equals the sum of the degrees on M_1 and M_2.

(2) If F_0 and F_1 are homotopic through F_t with $F_t(z) \neq 0$ throughout M, then F_0 and F_1 have the same degree. A particular case is

(3) *Rouché's Principle:* Let $H: M \to \mathbb{C}^n$ be a continuous map such that for every $x \in M$ there is an $1 \le i \le n$ with $|h_i(x)| < |f_i(x)|$, then F and $F+H$ have the same degree because F and $F+H$ are homotopic through $F_t = F + tH$. (The components of H and F are denoted by h_i and f_i respectively.)

(4) Let D be a compact region in $\mathbb{C}^n$ with smooth boundary ∂D. Let $F: D \to \mathbb{C}^n$ be continuous and $F(z) \neq 0$ throughout D. Then $F|\partial D$ has degree 0.

Let $F: (U,a) \to (\mathbb{C}^n, 0)$ be continuous with $F(z) \neq 0$ for $z \neq a$ in a neighbourhood U of a in $\mathbb{C}^n$. Let $\varepsilon > 0$ be so small that the closed ball $B_\varepsilon(a) = \{z \in \mathbb{C}^n: \|z-a\| \le \varepsilon\} \subset U$. Then F has the same degree on all spheres $S_\delta(a) = \partial B_\delta(a)$ for $0 < \delta \le \varepsilon$. This degree depends only on the germ of F at a. It is called the *multiplicity* $m_a(F)$.

<u>Lemma 1</u>: *Let* $D \subset \mathbb{C}^n$ *be a compact region with smooth boundary* ∂D . *Let* $F: D \to \mathbb{C}^n$ *be continuous with* $F(z) \neq 0$ *throughout* ∂D . *If* $F(z)=0$ *has finitely many solutions, the degree of* F *on* ∂D *equals the sum of the multiplicities* $m_x(F)$ *for all solutions* x *of* $F(z)=0$.

Proof: Remove a small open ε-ball about each x . The boundary of the remaining region D_o is the disjoint union of ∂D and of the spheres $S_\varepsilon(x)$. The orientation of $S_\varepsilon(x)$ as part of ∂D_o is opposite to the orientation as $\partial B_\varepsilon(x)$. The lemma follows from the properties (1) and (4) above, applied to ∂D_o.

<u>Lemma 2</u>: *Let* $F: (\mathbb{C}^n,a) \to (\mathbb{C}^n,0)$ *be a holomorphic germ, which is regular at* a . *Then* $m_a(F)=1$.

Proof: We have $F(z)=L(z-a)+h(z)$, where L is linear and $\lim \|h(z)\| / \|z-a\| = 0$ for $z \to a$. Since F is regular, $L \in GL(n,\mathbb{C})$. Therefore $\|h(z)\| < \|L(z-a)\|$ whenever $\|z-a\| = \varepsilon$, provided ε is sufficiently small. By Rouché F(z) and L(z-a) have the same multiplicity at a . Now L can be continuously transformed to the identity within $GL(n,\mathbb{C})$. Therefore F has the same multiplicity at a as $z \mapsto z-a$, viz. 1.

The following result justifies the name "multiplicity":

<u>Lemma 3</u>: *Let* $F: (U,a) \to (\mathbb{C}^n,0)$ *be a holomorphic map with* $F(z) \neq 0$ *for all* $z \neq a$ *in the neighbourhood* U *of* a *in* $\mathbb{C}^n$. *Let* ε *be so small that the closed ball* $B_\varepsilon(a) \subset U$, *and let* $c \in \mathbb{C}^n$ *be so small that* $\|c\| < \|F(z)\|$ *throughout* $S_\varepsilon(a)$. *Then* $F(z)=c$ *has* $m_a(F)$ *solutions within* $B_\varepsilon(a)$, *provided* F *is regular at every solution.*

Proof: The solutions of $F(z)=c$ form a discrete set because F is regular there. This set intersects $B_\varepsilon(a)$ in finitely many points. We have $F \neq 0$ throughout $S_\varepsilon(a)$. Then

$m_a(F)$ = degree of F on $S_\varepsilon(a)$, by definition
 = degree of F-c on $S_\varepsilon(a)$, by Rouché's Principle
 = sum of $m_z(F-c)$ for all $z \in B_\varepsilon(a)$ with $F(z)=c$, by lemma 1,
 = number of solutions of $F(z)=c$ within $B_\varepsilon(a)$ because $m_z(F-c)=1$ by lemma 2 .

The Nullstellensatz, more precisely III§8, theorem 3, implies that the holomorphic germ $F: (\mathbb{C}^n,a) \to (\mathbb{C}^n,0)$ is finite with respect to the ideal $<0> \subset {}_n\mathcal{O}_a$ if a is an isolated point of the fibre $F^{-1}(0)$. We apply further results of Chapter III: For dimension reasons F is also strict, compare III§9. By the summary at the end of III§7 there are neighbourhoods W of a and V of 0 such that $F: W \to V$ is surjective, proper, and has finite fibres. There is a holomorphic function $d \neq 0$ on V such that $F: W \to V$ is a smooth holomorphic covering with r sheets outside the subset $\{y: d(y)=0\}$ of V. For every $\varepsilon > 0$ there is a neighbourhood V_ε of 0 in V such that every $z \in W$ with $F(z) \in V_\varepsilon$ satisfies $\|z-a\| \leq \varepsilon$.

We combine these results with lemma 3: Let ε be so small that $B_\varepsilon(a) \subset W$, and choose some $c \in V_\varepsilon$ with $d(c) \neq 0$ such that $\|c\| < \|F(z)\|$ for all $z \in S_\varepsilon(a)$. Then $F(z)=c$ has r solutions in $B_\varepsilon(a)$, and F is regular at every solution. Therefore

$$\textit{number of sheets} = \textit{multiplicity}, \qquad r = m_a(F).$$

The summary at the end of III§7 contains further characterizations of the number r of sheets: The finite and strict germ F induces a finite ring extension $F^a: {}_n\mathcal{O} \hookrightarrow {}_n\mathcal{O}_a$. The induced extension of the quotient fields has degree

$$r = [{}_n\mathcal{M}_a : {}_n\mathcal{M}_o].$$

Let $<f_1,\ldots,f_n> \subset {}_n\mathcal{O}_a$ denote the ideal generated by the components f_i of F. The minimal number s of generators of ${}_n\mathcal{O}_a$ as ${}_n\mathcal{O}$-module via F^a is the dimension of the complex vector space ${}_n\mathcal{O}_a/<f_1,\ldots,f_n>$. In general $r \leq s$ with $r = s$ if and only if ${}_n\mathcal{O}_a$ is a free module. But this is the case, thanks to III§12, theorem 3. Thus the main result of this section is obtained.

Theorem: *For a finite holomorphic germ* $F: (\mathbb{C}^n,a) \to (\mathbb{C}^n,0)$ *with components* $f_1,\ldots,f_n$ *the multiplicity* $m_a(F)$ *equals the dimension of the complex vector space* ${}_n\mathcal{O}_a/<f_1,\ldots,f_n>$.

A more elementary proof of this theorem, which does not require the results of Chapter III has been given by

Kuschnirenko and can be found in the book by Arnold, Gusein-Zade, and Varchenko from p.84 on.

§8 Weighted Homogeneous Polynomials

Let $W(\omega_1,\ldots,\omega_n;d)$ denote the finite dimensional complex vector space of weighted homogeneous polynomials with respect to $(\omega_1,\ldots,\omega_n)$ of total weight d, compare p.123. The monomials $z_1^{\alpha_1}\ldots z_n^{\alpha_n}$ with $\omega_1\alpha_1+\ldots+\omega_n\alpha_n=d$ form a base. Let

$$W = W(\omega_1,\ldots,\omega_n;d_1\ldots,d_n)$$

denote the complex vector space of all maps $F\colon \mathbb{C}^n \to \mathbb{C}^n$ whose components f_i belong to $W(\omega_1,\ldots,\omega_n;d_i)$, thus $W = W(\omega_1,\ldots,\omega_n;d_1)\times\ldots\times W(\omega_1,\ldots,\omega_n;d_n)$. For every $F\in W$ we have $F(0)=0$. The germ of F at 0 is finite if and only if $F^{-1}(0)=\{0\}$. Let $m(F)$ denote the multiplicity of F at 0 and let $m(F)=\infty$ if F is not finite at 0.

Lemma 1: *If* $F\in W$ *is finite at* 0, *then* $m(G)=m(F)$ *for all* G *in a neighbourhood of* F *in* W.

The proof uses the quasi-norm σ and the quasi-homothety s of p.67: Let $\omega = \text{lcm}(\omega_1,\ldots,\omega_n)$ and let $\alpha_i=\omega/\omega_i$. Then

$$\sigma(z) = (|z_1|^{\alpha_1}+\ldots+|z_n|^{\alpha_n})^{1/\omega} \quad \text{for} \quad z\in\mathbb{C}^n$$

is the quasi-norm, and

$$s(t,z) = (t^{\omega_1}z_1,\ldots,t^{\omega_n}z_n) \qquad \text{for } t\in\mathbb{C} \text{ and } z\in\mathbb{C}^n$$

is the quasi-homothety. The set

$$S = \{z\in\mathbb{C}^n\colon \sigma(z) = 1\}$$

is compact; $0\notin S$. The continuous function $z\mapsto \max\{|f_i(z)|\colon i=1,\ldots,n\}$ has a minimum $c>0$ on S, i.e.

(1) *For every* $y \in S$ *there is an* i *such that* $|f_i(y)| \geq c$.

Let g_i denote the i-th component of $G \in W$. The function

$$\rho: W \to \mathbb{R},\ G \to \max\{|g_i(y)-f_i(y)| \ :\ y \in S \text{ and } i=1,\dots,n\}$$

is continuous. It has the value 0 for $G=F$. Thus for all G close to F the value $\rho(G) < c$, in other words: For every $y \in S$, for every $G \in W$ close to F, and for every $i=1,\dots,n$

(2) $$|g_i(y)-f_i(y)| < c \ .$$

Every $z \in \mathbb{C}^n \setminus \{0\}$ can uniquely be written as $z=s(t,y)$ where $t > 0$ and $y \in S$: Choose $t=\sigma(z)$ and $y_i = t^{-\omega_i} z_i$. Then $h(z) = t^d h(y)$ for every $h \in W(\omega_1,\dots,\omega_n;d)$. Therefore (1) and (2) imply:

(1') *For every* $z \in \mathbb{C}^n$ *there is an* i *such that* $|f_i(z)| > c \cdot \sigma(z)^{d_i}$.

(2') *For every* $z \in \mathbb{C}^n$, *for every* $G \in W$ *close to* F, *and for every* $i=1,\dots,n$

$$|g_i(z)-f_i(z)| < c \cdot \sigma(z)^{d_i} \ .$$

Together they imply: For every $z \in \mathbb{C}^n$ there is an i such that for every G close to F in W

$$|g_i(z)-f_i(z)| < |f_i(z)| \ .$$

Then Rouché's principle, compare §7(3), yields $m(G)=m(F)$.

<u>Theorem 2</u>: *For every* $F \in W=W(\omega_1,\dots,\omega_n;d_1,\dots,d_n)$ *either*

(3) $$m(F) = \frac{d_1 \cdots d_n}{\omega_1 \cdots \omega_n}$$

or $m(F) = \infty$. *The set* $\{F: m(F)=\infty\}$ *is Zariski closed in* W.

Proof: Let $W_k=\{F \in W: m(F) \geq k\}$. By lemma 1, $W_k \setminus W_{k+1}$ is open (possibly empty) for $k < \infty$. On the other hand W_k is Zariski closed in W, including the case $k=\infty$. This follows from the main result of §7, $m(F) = \dim \mathcal{O}/\langle f_1,\dots f_n\rangle$ and §4, proposition 1, applied to the n algebraic (even linear) families $\Phi_i: W \to \mathcal{O}$, $\Phi_i(F)$ = germ of f_i at 0. Let $\lambda = \min\{k: W_k \neq \emptyset\}$. Then $W_\lambda=W$ and $W_{\lambda+1}= W_\infty$.

It remains to calculate λ . First the case $(\omega_1,\dots,\omega_n) = (1,\dots,1)$ is considered. The map $H(z) = (z_1^{d_1},\dots,z_n^{d_n})$ belongs to $W(1,\dots,1;d_1,\dots,d_n)$, and

$m(H) = \dim \mathcal{O}/<z_1^{d_1},\dots,z_n^{d_n}> = d_1\dots d_n$. Thus (3) holds true in this case. If $F \in W(\omega_1,\dots,\omega_n;d_1,\dots,d_n)$, we compose with $G(z)=(z_1^{\omega_1},\dots,z_n^{\omega_n})$. Then F G $W(1,\dots,1;d_1,\dots d_n)$ and so $m(F\circ G) = d_1\dots d_n$ or $= \infty$. Since $m(F\circ G) = m(F)\cdot m(G)$ and $m(G) = \omega_1\dots\omega_n$, formula (3) follows.

The ideals in ${}_n\mathcal{O}$ generated by all monomials $z^{\alpha_1}..z^{\alpha_n}$ with $\Sigma\omega_i\alpha_i \geq d$ resp. $\Sigma\omega_i\alpha_i > d$ are denoted by

(4) $W^+(\omega_1,..,\omega_n;d)$ resp. $W^{++}(\omega_1,..,\omega_n;d)$.

If $h \in {}_n\mathcal{O}$ is written as convergent series $h = \sum_{r=o}^{\infty} {}_rh$ of polynomials ${}_rh \in W(\omega_1,..,\omega_n;r)$, then $h \in W^+(..;d)$ resp. $\in W^{++}(..;d)$ if and only if ${}_rh=0$ for $r<d$ resp. $r\leq d$. For every $h \in W^{++}(..;d)$ there are constants $C > 0$ and $\delta > 0$ such that

(5) $|h(z)| < C\cdot\sigma(z)^{d+1}$ for all $z \in \mathbf{C}^n$ with $\sigma(z) \leq \delta$.

Proof of (5): There is a convergent series $M = \Sigma M_r$ of real numbers $M_r \geq 0$ such that $|{}_rh(z)| < M_r$ for all z in a neighbourhood U of 0 . If δ is sufficiently small, the set $\{z: \sigma(z)\leq\delta\}$ is contained in U . Let $y=s(\delta/\sigma(z),z)$. Then $\sigma(y)=\delta$ and $z=s(t,y)$ with $t=\sigma(z)/\delta$. Further

$$h(z) = \sum_{r>d} {}_rh(z) = \sum_{r>d} t^r\, {}_rh(y) = t^{d+1} \sum_{r>d} t^{r-d-1}\, {}_rh(y) \text{ for } z \in U .$$

Let $t\leq 1$. Then $|h(z)| < t^{d+1} \sum_{r>d} |{}_rh(y)| < \sigma(z)^{d+1}M/\delta^{d+1}$. This implies (5) with $C = M/\delta^{d+1}$.

The vector spaces of all mapping germs $F: (\mathbf{C}^n,0) \to (\mathbf{C}^n,0)$ whose i-th components f_i belong to $W^+(\omega_1,..,\omega_n;d_i)$ resp. $W^{++}(\omega_1,..,\omega_n;d_i)$ are denoted by

(6) $W^+(\omega_1,..,\omega_n;d_1,..,d_n)$ resp. $W^{++}(\omega_1,..,\omega_n;d_1,..,d_n)$

Theorem 3: *Let* $F \in W(\omega_1,..,\omega_n;d_1,..,d_n)$ *be finite at* 0 *and let* $H \in W^{++}(\omega_1,..,\omega_n;d_1,..,d_n)$. *Then* F *and* F+H *have the same multiplicity at* 0 .

$$m(F+H) = m(F) = \frac{d_1 \ldots d_n}{\omega_1 \ldots \omega_n} \ .$$

Proof: We apply (5) to the components h_i of H: There are constants $C > 0$ and $\delta > 0$ such that

(7) $\quad |h_i(z)| < C\cdot\sigma(z)^{d_i+1}$ for all $z \in \mathbb{C}^n$ with $\sigma(z)<\delta$.

On the other hand, using (1') above, there is a constant $c>0$ such that for every $z \in \mathbb{C}^n$ there is an i with

(8) $$|f_i(z)| > c\,\sigma(z)^{d_i} \ .$$

The set $V = \{z \in \mathbb{C}^n : \sigma(z) < \min\{\delta, c/C\}\}$ is a neighbourhood of 0 in $\mathbb{C}^n$. For every $z \in V$ there is an i such that

$$|f_i(z)| > |h_i(z)|$$

because of (7) and (8). Then $m(F+H)=m(F)$ follows from Rouché's principle.

If $f \in W(\omega_1,\ldots,\omega_n;d)$, then the map grad$f$ with components $\partial f/\partial z_i$ belongs to $W(\omega_1,\ldots,\omega_n;d-\omega_1,\ldots,d-\omega_n)$. The same holds true for W^+ instead of W . Therefore the theorems 2 and 3 yield

<u>Theorem 2'</u>: *For every* $f \in W(\omega_1,\ldots,\omega_n;d)$ *either* $\mu(f) = \left(\frac{d}{\omega_1} - 1\right)\ldots\left(\frac{d}{\omega_n} - 1\right)$ *or* $\mu(f) = \infty$. *The set* $\{f : \mu(f)=\infty\}$ *is Zariski closed in* W .

Every $f \in W^+ = W^+(\omega_1,\ldots,\omega_n;d)$ can uniquely be decomposed into $f=f_0+f_1$ with $f_0 \in W$ and $f_1 \in W^{++}$. Then f_0 is called the *principal part* of f . The principal part depends not only on f but also on $\omega_1,\ldots,\omega_n;d$.

<u>Theorem 3'</u>: *If* $f \in W^+(\omega_1,\ldots,\omega_n;d)$ *and if for the principal*

part $\mu(f_0) < \infty$, *then* $\mu(f) = \mu(f_0) = \left(\frac{d}{\omega_1} - 1\right) \dots \left(\frac{d}{\omega_n} - 1\right)$.

We are particularly interested in the subsets

(10) $\quad \mathring{W}(\omega_1,\dots,\omega_n;d) = \{f \in W(\omega_1,\dots,\omega_n;d) : \mu(f) < \infty\}$ and

(11) $\quad \mathring{W}^+(\omega_1,\dots,\omega_n;d) = \{f \in W^+(\omega_1,\dots,\omega_n;d) : \mu(f_o) < \infty\}$.

The last statement of theorem 2' implies that $\mathring{W}^+$ is open and dense in W^+ with respect to the topology of W^+ as subspace of $\mathcal{O}$.

Proposition 4: *If* W^+ *contains a simple germ, then all germs in* $\mathring{W}^+$ *are simple and mutually holomorphically equivalent.*

Proof: If a (μ=constant)-family has a simple member, then all its members are mutually equivalent, see §6, proposition 4. We apply this to the 1-parameter family $(1-t)f+tf_o$ for $f \in \mathring{W}^+$ with principal part f_0 , and to $\mathring{W}$ considered as family. In both cases μ=constant because of theorems 3' and 2' . Thus it suffices to find a simple $f \in \mathring{W}^+$. There is a simple $g \in W^+$. All germs in a neighbourhood of g are also simple. Since $\mathring{W}^+$ is dense in W^+ , there is a simple $f \in \mathring{W}^+$.

Proposition 5: *Let* $f \in \mathring{W}^+$. *If there is a* $u \in W^+ \setminus \mathfrak{m}\Delta(f)$, *then* f *is not simple and* W^+ *does not contain any simple germ.*

Proof: Consider the 1-parameter family $f_t = t+tu$ in W^+ for $t \in \mathbb{C}$. There is a neighbourhood T of 0 in $\mathbb{C}$, such that $t \in T$ implies $f_t \in \mathring{W}^+$, because $\mathring{W}^+$ is open in W^+. Therefore $\mu(f_t) = \mu(f)$ for $t \in T$ by theorem 3'. If f were simple, $f_t \sim f$ by §6, proposition 4. The family is miniverse because $u \in \mathfrak{m}\Delta(f)$. By §6, proposition 1 there is a neighbourhood T_0 of 0 in $\mathbb{C}$ such that $f_t \not\sim f$ for $t \in T_0 \setminus \{0\}$. So f cannot be simple.

Examples: The following germs f are not simple because $u \in W^+ \setminus \mathfrak{m}\Delta(f)$.

f	$x^3+y^3+z^3$	$x^4+y^4+z^2$	$x^6+y^6+z^2$
u	xyz	x^2y^2	x^4y

<u>Remark concerning the notation</u>: Obviously $W(\omega_1,\dots,\omega_n;d) = W(q\omega_1,\dots,q\omega_n;qd)$. In the following it will be convenient to admit *rational* $\omega_1,\dots,\omega_n$. This allows the normalization $d=1$. The common Milnor number of the germs in $\mathring{W}^+(\omega_1,\dots,\omega_n;1)$ is $\left(\frac{1}{\omega_1}-1\right)\dots\left(\frac{1}{\omega_n}-1\right)$. If this number is not a positive integer, $\mathring{W}^+(\omega_1,\dots,\omega_n;1)$ is empty.

§9 The Classification of Holomorphic Germs

Arbitrary germs $f\in {}_3\mathfrak{m}$ with $k=\mu(f)<\infty$ are transformed into "normal forms" which are elements of W^+ or even $\mathring{W}$ for a few series of weights $(\omega_1,\omega_2,\omega_3;d)$. Let $f=\sum_{d=1}^{\infty} f_d$ denote the Taylor series with homogeneous polynomials f_d of degree d. There are two possibilities

(1) $f_1 \neq 0$ (2) $f_1 = 0$.

In case (1) f is regular, and all regular germs are mutually holomorphically equivalent, compare §2. In case (2) $f\in\mathfrak{m}^2$. There are three possibilities:

(2a) $f_2 = 0$ (2b) f_2 is not a complete square.

(3) f_2 is a complete square $\neq 0$.

In case (2a) $f\in W^+(1,1,1;3)$. Now $x^3+y^3+z^3\in\mathring{W}(1,1,1;3)$ is not simple, and so f is not simple, compare the end of §8. The case (2b) will be investigated in §10 and §11 below. The result will be:

In case (2b) $f\sim x^{k+1}+y^2+z^2\in\mathring{W}(2,k+1,k+1;2(k+1))$ *for* $k=1,2,\dots$

In case (3) a linear coordinate transformation yields $f=z^2$ mod $\mathfrak{m}^3$, and so $f\in W^+(2,2,3;6)$. The principal part $f_0\in W(2,2,3;6)$ has the form $f_0=\varphi_3(x,y)+z^2$ where $\varphi_3(x,y)=$

$f_0(x,y,0)$ is a cubic form. There are three possibilities:

(3a) $\varphi_3 = 0$ (3b) φ_3 is not a complete cube.

(4) φ_3 is a complete cube $\neq 0$.

In case (3a) $f \in W^+(1,1,2;4)$. Now $x^4+y^4+z^2 \in \mathring{W}(1,2,2;4)$ is not simple, and so f is not simple, compare again the end of §8. The case (3b) will be investigated in §10 and §11 below. The result will be:

In case (3b) $f \sim x^{k-1}+xy^2+z^2 \in \mathring{W}(2,k-2,k-1;2(k-1))$, $k=4,5,\ldots$

In case (4) a linear coordinate transformation of (x,y) yields $f=z^2+y^3+azx^2+bzy^2+cz^2x+dz^2y \bmod \mathfrak{m}^4$, and so $f \in W^+(3,4,6;12) = W^+(1/4,1/3,1/2;1)$. We consider also $W^+(1/p,1/3,1/2;1)$ for rational numbers $p > 4$. In §10 and §11 below we shall see that there are four possibilities:

(4a) $f \sim x^4+y^3+z^2 \in \mathring{W}(1/4,1/3,1/2;1)$

(4b) $f \sim x^3y+y^3+z^2 \in \mathring{W}(2/9,1/3,1/2;1)$

(4c) $f \sim x^5+y^3+z^2 \in \mathring{W}(1/5,1/3,1/2;1)$

(4d) $f \sim g \in W^+(1/6,1/3,1/2;1)$.

Now $x^6+y^3+z^2 \in \mathring{W}(1/6,1/3,1/3;1)$ is not simple, and so f is not simple in case (4d) , compare again the end of §8.

In the cases (1a), (2b), (3b) and (4a-c) the germ f is simple as will be shown in §11 below. Let

$$A_k = \{f: f \sim x^{k+1}+y^2+z^2\} \quad k=0,1,2,\ldots$$
$$D_k = \{f: f \sim x^{k-1}+xy^2+z^2\} \quad k=4,5,\ldots$$
$$E_6 = \{f: f \sim x^4+y^3+z^2\}$$
$$E_7 = \{f: f \sim x^3y+y^3+z^2\}$$
$$E_8 = \{f: f \sim x^5+y^3+z^2\} .$$

The index denotes the Milnor number. The following theorem summarizes the results of this and the following two sections:

<u>Theorem</u>: *The set of simple germs in* $_3\mathfrak{m}$ *is the disjoint union of the holomorphic equivalence classes* ($\mathcal{G}$*-orbits*) A_k

for $k=0,1,2,\ldots$, D_k *for* $k=4,5,\ldots$, *and* E_6, E_7, E_8 . *The set of the non simple germs in* $_3\mathfrak{m}$ *is the disjoint union of the following three sets* U,V,W:

$U = W^+(1/3,1/3,1/3;1) = {}_3\mathfrak{m}^3$,

$V = \{f: f \sim g \in W^+(1/4,1/4,1/2;1)$ with $g(0,0,z)=z^2 \bmod {}_3\mathfrak{m}^3\}$,

$W = \{f: f \sim g \in W^+(1/6,1/3,1/2;1)$ with $g(0,0,z)=z^2 \bmod {}_3\mathfrak{m}^3$,

and $g(0,y,0)=y^3 \bmod {}_3\mathfrak{m}^4\}$.

We have $\mu(f) \geq 8$ *for* $f \in U$, $\mu(f) \geq 9$ *for* $f \in V$, *and* $\mu(f) \geq 10$ *for* $f \in W$.

§10 Three Series of Holomorphic Germs

According to §9 three types of holomorphic germs $f \in {}_3\mathcal{O}$ need further investigations.

1st case, §9(2b):

$f \in W^+(\frac{1}{2},\frac{1}{2},\frac{1}{2};1)$; the principal part $f_o \in W(\frac{1}{2},\frac{1}{2},\frac{1}{2};1)$ is a quadratic form in three variables, which is not the square of a linear form. A linear coordinate transformation yields

$$f_o = ax^2+y^2+z^2 \quad , \quad a \in \mathbb{C} .$$

2nd case, §9(3b):

$f \in W^+(\frac{1}{3},\frac{1}{3},\frac{1}{2};1)$; the principal part $f_o \in W(\frac{1}{3},\frac{1}{3},\frac{1}{2};1)$ has the form $f_o = \varphi(x,y)+z^2$; here φ is a cubic form, which is not the cube of a linear form. A cubic form in two variables is the product of three linear forms, $\varphi = \lambda_1\lambda_2\lambda_3$. Since φ is not a cube, we may assume that λ_1 and λ_2 are linearly independent. Then after a linear coordinate transformation $x = \lambda_1$ and $y = \lambda_2$, and so $\varphi = xy(ax+by)$ with $a \neq 0$ or $b \neq 0$. We may assume that $b \neq 0$. Then after a suitable homothety $x \mapsto \alpha x$, $y \mapsto \beta y$ we obtain $\varphi = xy(cx+y)$. Thus the principal part of f becomes

$$f_o = cx^2y + xy^2 + z^2 \;, \quad c \in \mathbb{C} \;.$$

3rd case, §9(4):

$f \in W^+(\frac{1}{4},\frac{1}{3},\frac{1}{2};1)$; the principal part $f_o \in W(\frac{1}{4},\frac{1}{3},\frac{1}{2};1)$ has the form

(1) $$f_o = ax^4+2bx^2z+y^3+z^2 \;, \quad a,b \in \mathbb{C} \;.$$

<u>1st case:</u> We consider $W_p^+ = W_p^+(\frac{1}{p},\frac{1}{2},\frac{1}{2};1)$ for rational $p \geq 2$. Then $W_p^+ \supset W_q^+$ for $p \leq q$. We ask for those p such that $W_p^+ \neq W_q^+$ for $p < q$. Then the equation $\alpha/p+\beta/2+\gamma/2=1$ must have a solution (α,β,γ) with integers $\alpha \geq 1$, $\beta \geq 0$, $\gamma \geq 0$. This is the case if and only if p is an integer. Therefore we consider the inclusions

$$W_2^+ \supset W_3^+ \supset \ldots \quad \text{with} \quad \mu(f) > p-1 \quad \text{for} \quad f \in W_p^+ \;.$$

<u>Proposition 1:</u> *Let* f *with* $k=\mu(f)<\infty$ *belong to the first case. Then* $f \sim g \in \mathring{W}_{k+1}^+$ *for some* g.

Proof: Let $l = \max\{p\colon f \sim g \in W_p^+\}$. This maximum exists because $k \geq p-1$. We replace f by $g \in W_l^+$ and thus may assume $f \in W_l^+$. The principal part $f_o \in W_l$ has the following form

$$f_o = ax^l+y^2+z^2 \quad \text{for odd } l \;,$$

$$f_o = ax^2+2bx^my+2cx^mz+y^2+z^2 \quad \text{for even } l=2m \;.$$

If f_o has an isolated critical point, then $f \in \mathring{W}_l^+$ and $l=k+1$. If f_o has not an isolated critical point, then

$$f_o = y^2+z^2 \quad \text{for odd } l$$

$$f_o = (y+bx^m)^2 + (z+cx^m)^2 \quad \text{for even } l=2m \;.$$

This contradicts the maximality of l: We have $f=f_o+f_1$ with $f_1 \in W_l^{++} \subset W_{l+1}^+$. If l is odd, we have $f_o \in W_{l+1}^+$, and thus $f \in W_{l+1}^+$. If $l=2m$ is even, we transform by $y \mapsto y-bx^m$, $z \mapsto z-cx^m$. Then f_o becomes $y^2+z^2 \in W_{l+1}$ and again $f \in W_{l+1}^+$.

<u>2nd case</u>: We consider $W_p^+ = W^+(\frac{1}{p}\ \frac{p-1}{2p},\frac{1}{2};1)$ for rational $p\geq 3$. Again only integer p's must be considered:

$$W_3^+ \supset W_4^+ \supset \dots \text{ with } \mu(f) \geq p+1 \text{ for } f\in W_p^+ .$$

<u>Proposition 2</u>: *Let* f *with* $k=\mu(f)<\infty$ *belong to the second case. Then* $f\sim g\in \mathring{W}_{k-1}^+$ *for some* g .

The proof, which is quite similar to the proof of proposition 1, is left to the reader.

<u>3rd case</u>: We consider $W_p^+ = W^+(\frac{1}{p},\frac{1}{3},\frac{1}{2};1)$ for rational $p\geq 4$. Here $W_4^+ \supset W_{9/2}^+ \supset W_5^+ \supset W_6^+$ are the relevant inclusions. We stop at W_6^+ because W_6^+ contains no simple germs, compare §9(4d) . The Milnor numbers satisfy

$\mu(f)\geq 6$ for $f\in W_4^+$, $\mu(f)\geq 7$ for $f\in W_{9/2}^+$, $\mu(f)\geq 8$ for $f\in W_5^+$ and $\mu(f)\geq 10$ for $f\in W_6^+$.

<u>Proposition 3</u>: *Let* f *with* $k=\mu(f)<\infty$ *belong to the third case. Then* $f\sim g\in \mathring{W}_{(k/2)+1}^+$ *for* k=6,7,8 *or* $f\sim g\in W_6^+$ *for* $k>8$ *for some* g .

Proof: If the principal part f_0 of f , see (1), has an isolated critical point, then $f\in \mathring{W}_4^+$. If not, then $f_0= y^3+(bx^2+z)^2$. We have $f = f_0+f_1$ with $f_1\in W_4^{++}\subset W_{9/2}^+$. We transform by $z \mapsto z-bx^2$. Then f_0 becomes y^3+z^2 , and f_1 stays in $W_{9/2}^+$, thus $f\sim g\in W_{9/2}^+$. The germ $g\in W_{9/2}^+$ has principal part

$$g_o= cx^3y+y^3+z^2 \in W_{p/2} .$$

If $c \neq 0$, the critical point of g_o is isolated, and so $f\sim g\in \mathring{W}_{9/2}^+$. If c=0 , then $g\in W_5^+$. Then the new principal part is

$$g_o= dx^5+y^3+z^2 \in W_5 .$$

If $d \neq 0$, the critical point of g_0 is isolated and so $f \sim g \in \mathring{W}_5^+$. If $d = 0$, then $f \sim g \in W_6^+$.

§11 Simple Singularities

For each set $\mathring{W}_p^+$ of the last proposition a representative $\varphi \in \mathring{W}_p^+$ is chosen, see the following table:

φ	weights	$\varphi\mathcal{G}$	$\mu(\varphi)$	
$x^{k+1}+y^2+z^2$	$\frac{1}{k+1}, \frac{1}{2}, \frac{1}{2}; 1$	A_k	k	k=0,1,2,...
$x^{k-1}+xy^2+z^2$	$\frac{1}{k-1}, \frac{k-2}{2(k-1)}, \frac{1}{2}; 1$	D_k	k	k=4,5,...
$x^4+y^3+z^2$	$\frac{1}{4}, \frac{1}{3} \frac{1}{2}; 1$	E_6	6	
$x^3y+y^3+z^2$	$\frac{2}{9}, \frac{1}{3}, \frac{1}{2}; 1$	E_7	7	
$x^5+y^3+z^2$	$\frac{1}{5}, \frac{1}{3}, \frac{1}{2}; 1$	E_8	8	

If $\varphi \in \mathring{W}_p$ is simple, then all germs $f \in \mathring{W}_p^+$ are holomorphically equivalent to φ because of §8, proposition 4. From §9, and §10 we know already that among all orbits $f\mathcal{G} \subset {}_3\mathfrak{m}$ only A_k, D_k, E_6, E_7, and E_8 have a chance to be simple, furthermore that these orbits are distinct. It remains to show that the germs φ of the table are actually simple. This result is contained in the following

<u>Proposition:</u> *The union* $A_o \cup \ldots \cup A_k$ *is a neighbourhood of* A_k *in* $\mathfrak{m}$, *the union* $A_o \cup \ldots \cup A_{k-1} \cup D_4 \ldots \cup D_k$ *is a neighbourhood of* D_k *in* $\mathfrak{m}$, *the union* $A_o \cup \ldots \cup A_{k-1} \cup D_4 \ldots \cup D_{k-1} \cup E_6 \cup \ldots \cup E_k$ *is a neighbourhood of* E_k *in* $\mathfrak{m}$.

The proof relies on §6, Corollary 3: We shall calculate universal unfoldings of each φ of the table and shall show:

(I) For $\varphi \in A_k$ all germs of the unfolding belong to $A_o \cup \ldots \cup A_k$.

(II) For $\varphi \in D_k$ all germs of the unfolding belong to A_l for some l or to $D_4 \cup \ldots \cup D_k$.

(III) For $\varphi \in E_k$ all germs of the unfolding belong to A_l or D_l for some l or to $E_6 \cup \ldots \cup E_k$.

The semi-continuity of the Milnor number implies $1 \leq k$. Therefore φ is simple. Using §5, proposition 5 A_k can be dropped from the neighbourhood of D_k, and both A_k and D_k can be dropped from the neighbourhood of E_k .

(I) implies $f \in A_k$ for every germ f with $\mu(f)=k < \infty$, which belongs to the first case of §10. This will be used for the proofs of (II) and (III).

(II) implies $f \in D_k$ for every germ f with $\mu(f)=k < \infty$, which belongs to the second case of §10. This will be used for the proof of (III).

The proofs of (I)-(III) proceed by induction on k .

Proof of (I): The germs of A_o are regular, thus $A_o = \mathfrak{m} \setminus \mathfrak{m}^2$ is open in $\mathfrak{m}$. The induction hypothesis "A_l is simple for $l<k$" implies $\mathring{W}^+_p = \mathring{W}^+(1/p,1/2,1/2;1) \subset A_{p-1}$ for $p \leq k$. A universe unfolding of $x^{k+1}+y^2+z^2$ consists of the germs

$$(A_k) \qquad g = x^{k+1}+y^2+z^2+\sum_{j=1}^{k} t_j x^j+uy+vz \ , \quad t_j,u,v \in \mathbb{C} \ .$$

If $(t_1,u,v) \neq (0,0,0)$, then g is regular. If $u=v=t_1=\ldots=t_{p-1}=0$ and $t_p \neq 0$ for some $1 \leq p \leq k$, then $g \in \mathring{W}^+_p \subset A_{p-1}$ by induction hypothesis. If $u=v=t_1=\ldots=t_k=0$, then $g=x^{k+1}+y^2+z^2 \in A_k$.

Proof of (II): A universe unfolding is

$$(D_k) \qquad g = x^{k-1}+xy^2+z^2+\sum_{j=1}^{k-2} t_j x^j+uy+vz+w_1xy+w_2y^2 \ ,$$

$$t_j,u,v,w_1,w_2 \in \mathbb{C} \ .$$

1st case: If $(t_1,u,v) \neq (0,0,0)$, then g is regular.

2nd case: If $t_1=u=v=0$, but $(t_2,w_1,w_2) \neq (0,0,0)$, then g has the principal part $g_o = z^2+t_2x^2+w_1xy+w_2y^2 \in W(1/2,1/2,1/2;1)$. This is not the square of a linear form. Thus g belongs to the first case of §10, and so $g \in A_l$ for some l .

3rd case (for $k \geq 5$ only): If $u=v=w_1=w_2=t_1=t_2=\ldots=t_{p-2}=0$

$t_{p-1} \neq 0$ for some $p < k$, then $g \in \mathring{W}^+_{p-1} =$
$\mathring{W}^+(\frac{1}{p-1}, \frac{p-2}{2(p-1)}, \frac{1}{2}; 1)$.

4th case: If $u=w=w_1=w_2=t_1=\ldots=t_{k-2}=0$, then $g = x^{k-1}+xy^2+z^2 \in D_k$.

The induction begins with $k=4$. Let g belong to the unfolding (D_4). In the first case $g \in A_0$, in the second case $g \in A_1$, the third case does not occur, in the fourth case $g = \varphi \in D_4$. The induction hypothesis "D_p is simple for $p<k$" implies $\mathring{W}^+_{p-1} \subset D_p$ for every $p<k$. Let g belong to the unfolding (D_k). In the first two cases $g \in A_1$, in the third case $g \in D_p$ for some $p<k$ using the induction hypothesis, in the fourth case $g = \varphi \in D_k$.

Proof of(III): The corresponding universe unfoldings are

(E_6) $g = x^4+y^3+z^2+t_1x^3+t_2x^2y+ \varphi(x,y)+\lambda(x,y,z)$

(E_7) $g = x^3y+y^3+z^2+t_1x^3+t_2x^2y+t_3xy^2+ \varphi(x,y)+\lambda(x,y,z)$

(E_8) $g = x^5+y^3+z^2+t_1x^4+t_2x^3y+u_1x^3+u_2x^2y+ \varphi(x,y)+\lambda(x,y,z)$.

Here λ is a linear form and φ is a quadratic form. If $\lambda \neq 0$, then $g \in A_0$. If $\lambda = 0$, but $\varphi \neq 0$, then the total quadratic form $z^2+\varphi(x,y)$ of g is not the square of a linear form. Thus g belongs to the 1st case of §10 and so $g \in A_1$. There remain the cases $\lambda = \varphi = 0$: With (E_6) either $t_1=t_2=0$ or the total cubic form $y^3+t_1x^3+t_2x^2y$ of g is not the cube of a linear form. Then g belongs to the 2nd case of §10 and so $g \in D_1$. This completes the proof for E_6. We note for later use the consequence

$$\mathring{W}^+(1/4,1/3,1/2;1) \subset E_6 .$$

With (E_7) either $g \in D_1$ or the total cubic form of g is the cube of a linear form, i.e.

$$y^3+t_1x^3+t_2x^2y+t_3xy^2 = (y+\alpha x)^3 .$$

Then we transform by $y \mapsto (y-\alpha x)$ and obtain $g = x^3(y-\alpha x)+y^3+z^2$. If $\alpha \neq 0$, $g \in \mathring{W}(1/4,1/3,1/2;1) \subset E_6$. If $\alpha=0$, $g = \varphi \in E_7$. This completes the proof for E_7. We note also

$$\mathring{W}^{+}(2/9,1/3,1/2;1) \subset E_7 .$$

With (E_8) either $g \in D_1$ or $u_1=u_2=0$. Then $g \in \mathring{W}^{+}(2/9,1/3,1/2;1) \subset E_7$ if $t_2 \neq 0$ or $g \in \mathring{W}^{+}(1/4,1/3,1/2;1) \subset E_6$ if $t_2=0, t_1 \neq 0$ or $g=\varphi \in E_8$ if $t_1=t_2=0$. This completes the proof for E_8 .

§12 Adjacency

In this section all adjacency relations between the simple germs in $\mathfrak{m}_3$ are established. Adjacency is a relation between the $\mathcal{G}$-orbits (=holomorphic equivalence classes). Let X denote a simple orbit, and let Y denote another orbit such that $Y \prec X$ and $Y \neq X$. The following table contains all possibilities

X	Y with $Y \prec X$, $Y \neq X$
A_k	A_1 for $1 < k$
D_k	A_1 and D_1 for $1 < k$
E_k	A_1, D_1, and E_1 for $1 < k$

The results of this table are partially contained in the proposition of the preceding section. The following adjacencies remain to be established:

$$A_{k-1} \prec A_k \qquad A_{k-1} \prec D_k$$
$$D_{k-1} \prec D_k \qquad D_{k-1} \prec E_k$$
$$E_{k-1} \prec E_k \qquad A_{k-1} \prec E_k \quad .$$

In order to show $Y \prec X$ we choose some $f \in X$ and specify an unfolding f_t of $f=f_o$ such that there exist arbitrary small t with $f_t \in Y$.

$A_k \succ A_{k-1}$:
$f_t = x^{k+1}+y^2+z^2+tx^k$. If $t\neq 0$, then $f_t \in \mathring{W}^+(1/k,1/2,1/2;1) \subset A_{k-1}$.

$D_k \succ D_{k-1}$:
$f_t = x^{k-1}+xy^2+z^2+tx^{k-2}$. If $t\neq 0$, then $f_t \in \mathring{W}^+(\frac{1}{k-2}, \frac{k-3}{2(k-2)}, \frac{1}{2};1)$ $\subset D_{k-1}$.

$E_7 \succ E_6$:
$f_t = x^3y+y^3+z^2+tx^4$. If $t\neq 0$, then $f_t \in \mathring{W}^+(1/4,1/3,1/2;1) \subset E_6$.

$E_8 \succ E_7$:
$f_t = x^5+y^3+z^2+tx^3y$. If $t\neq 0$, then $f_t \in \mathring{W}^+(2/9,1/3,1/2;1) \subset E_7$.

$D_{2n+2} \succ A_{2n+1}$:
$f_t = x^{2n+1}-xy^2+z^2+t(x^n+y)^2$. Transform by $y \mapsto y-x^n$ (with x and z fixed). Then $f_t \sim x^{2n+1}-x(y-x^n)^2+ty^2+z^2 = 2x^{n+1}y+ty^2+z^2-xy^2$ $\in \mathring{W}^+(1/(2n+2),1/2,1/2;1) \subset A_{2n+1}$ for $t\neq 0$.

$D_{2n+1} \succ A_{2n}$:
$f_t = x^{2n}+xy^2+z^2+(2x^n+ty)ty = xy^2+z^2+(x^n+ty)^2$. Let $t\neq 0$. Choose u such that $u^n=t$ and transform by $x \mapsto ux$, $y \mapsto y-x^n$. Then $f_t \sim ux(y-x^n)^2+t^2y^2+z^2 = ux^{2n+1}+t^2y^2+z^2+uxy^2-2ux^{n+1}y \in$ $\mathring{W}^+(1/(2n+1),1/2,1/2;1) \subset A_{2n}$.

$E_6 \succ D_5$:
$f_t = x^4+y^3+z^2+txy^2$. If $t\neq 0$, then $f_t \in \mathring{W}^+(1/4,3/8,1/2;1) \subset D_5$.

$E_7 \succ D_6$:
$f_t = x^3y+y^3+z^2+txy^2$. If $t\neq 0$, then $f_t \in \mathring{W}^+(1/5,2/5,1/2;1) \subset D_6$.

$E_8 \succ D_7$:
$f_t = x^5+y^3+z^2+(2x^2+ty)txy = y^3+x(x^2+ty)^2+z^2$. Let $t\neq 0$. Transform by $x \mapsto ux$ where $u^2=t$, and $y \to y-x^2$. Then $f_t \sim (y-x^2)^3+u^5xy^2+z^2 = -x^6+u^5xy^2+z^2+y^3+3yx^4-3y^2x^2 \in$ $\mathring{W}^+(1/6,5/12,1/2;1) \subset D_7$.

$E_6 \succ A_5$:

$f_t = x^4+y^3+z^2+(2x^2+ty)ty = y^3+(x^2+ty)^2+z^2$. Let $t \neq 0$. Transform by $x \mapsto ux$ where $u^2=t$, $y \mapsto y-x^2$. Then $f_t \sim (y-x^2)^3+t^2y^2+z^2 = -x^6+t^2y^2+z^2+y^3-3y^2x^2+3yx^4 \in \overset{\circ}{W}{}^+(1/6,1/2,1/2;1) \subset A_5$.

The last two cases are more complicated. Let $W_p = W(1/p,1/2,1/2;1)$.

$E_7 \succ A_6$:

We begin with a three parameter unfolding, $t=(a,b,c)$, $f_t = x^3y+y^3+z^2+a(x^2+by)^2+cxy^2$. If $ab \neq 0$ then f_t belongs to the first case of §10 and thus $f_t \in A_k$ for some $k \leq 6$. We want to obtain $k=6$. Transform by $x \mapsto bx$, $y \mapsto y-bx^2$. Then $f_t \sim b^3(c-b)x^5-b^3x^6+b^2(b-2c)x^3y+ab^2y^2+z^2+g$, $g \in W_6^{++}$. The first condition

(1) $$b = c$$

destroys the coefficient of x^5 . Then $f_t \sim b^2(-bx^6-bx^3y+ay^2)+z^2+g$. The second condition

(2) $$b = -4a$$

yields $f_t \sim 64a^3(x^3+y/2)^2+z^2+g$. The transformation $y \mapsto 2(y-x^3)$ maps g into some $h \in W_6^{++} \subset W_7^+$, and f_t becomes equivalent to $64a^3y^2+z^2+h \in W_7^+$, and so $\mu(f_t) \geq 6$. Both conditions can obviously be satisfied by arbitrary small $t \neq 0$, viz. $t = (a,-4a,-4a)$ and $a \neq 0$.

$E_8 \succ A_7$:

The procedure of the previous case with W_7^{++} instead of W_6^{++} is not quite fitting because $y \mapsto y-x^3$ does not map W_7^{++} into itself. Rather W_6^{++} in the previous case must be replaced by $W^{++}(1/7,3/7,1/2;1)$ which is mapped into $W_7^{++} =$

$W^{++}(1/7,1/2,1/2;1)$ by the transformation $y \mapsto y-x^3$.

We begin with a four parameter unfolding, $t=(a,b,c,d)$, $f_t = x^5+y^3+z^2+a(x^2+by)^2+cxy^2+dx^3y$. If $ab\neq 0$, then $f_t \in A_k$ for some $k \leq 7$. We transform by $x \mapsto bx$, $y \mapsto y-bx^2$. Then $f_t \sim b^3(b^2+c-bd)x^5+3b^4x^4y-b^3x^6+z^2+ab^2y^2+bcxy^2+b^2(bd-2c)x^3y+g$, $g \in W^{++}(1/7,3/7,1/2;1)$. The coefficient of x^5 is destroyed by the condition

(3) $$b^2+c = bd .$$

Then $f_t \sim -b^2(bx^6+(c-b^2)x^3y-ay^2)+3b^4x^4y+bcxy^2+z^2+g$. The second condition is

(4) $$(b^2-c)^2+4ab = 0 .$$

There are α,β with $\alpha^2=a$, $\beta^2=-b$, $2\alpha\beta=b^2-c$ and so $f_t \sim \beta^4(\beta x^3+\alpha y)^2+3\beta^8x^4y+\beta^3(2\alpha-\beta^3)xy^2+z^2+g$. The homothety $x \mapsto x/\beta$, $y \mapsto y/\beta$ simplifies this to $f_t \sim (x^3+\alpha\beta y)^3+3\beta^3x^4y+(2\alpha-\beta^3)xy^2+z^2+g'$, $g' \in W^{++}(1/7,3/7,1/2;1)$. Let $\gamma^2=\alpha\beta$. Transform by $x \mapsto \gamma x$, $y \mapsto y-\gamma x^3$. Then $f_t \sim 2\gamma^5(\alpha-2\beta^3)x^7+h$, $h \in W_7^{++} \subset W_8^+$. The last condition

(5) $$\alpha = 2\beta^3$$

yields $f_t \in W_8^+$, and so $\mu(f_t) \geq 7$. The three conditions (3), (4), and (5) can be satisfied by arbitrary small $t=(a,b,c,d)$: Let $\beta \in \mathbb{C}\setminus\{0\}$ and choose $a=4\beta^6$, $b=-\beta^2$, $c=-3\beta^4$, $d=2\beta^2$.

§ 13 Conclusion and Outlook

Earlier in II§8, table 4, and in IV§3 the finite subgroups G of $SL(2,\mathbb{C})$ and the corresponding quotient singularities $\mathbb{C}^2/G$ have been related to polynomials $\varphi(x,y,z)$. These polynomials turn out to form a complete set of representatives of the simple orbits A_{n-1}, D_{q+1}, E_6, E_7, E_8 : Obvious coordinate transformations transform the polynomials φ of loc. cit. into the representatives which have been used in §9 for the definition of A_k, ... This observation together with IV§3, proposition 4, leads to the

Conclusion: *A germ $f \in {}_3\mathfrak{m}$ with an isolated critical point is simple if and only if its zero set $N(f)$ is isomorphic to the germ of the quotient variety $\mathbb{C}^2/G$ of a finite subgroup $G < SL(2,\mathbb{C})$. This relation $f \leftrightarrow G$ establishes a one-to-one correspondence between the simple germs of holomorphic functions f in three variables up to holomorphic equivalence (simple orbits in ${}_3\mathfrak{m}$) and the finite subgroups $G < SL(2,\mathbb{C})$ up to conjugacy, given by the following table:*

Simple orbit	representing function	type of $G < SL(2,\mathbb{C})$
A_k	$x^{k+1} + y^2 + z^2$	cyclic of order k+1
D_k	$x^{k-1} + xy^2 + z^2$	binary (k-2)-dihedral
E_6	$x^4 + y^3 + z^2$	binary tetrahedral
E_7	$x^3y + y^3 + z^2$	binary octahedral
E_8	$x^5 + y^3 + z^2$	binary icosahedral

Outlook: An important aspect of the investigation of isolated singularities remains beyond the scope of this book: The fibration viewpoint. Here Milnor's book (1) of 1968 has been seminal. Husein-Zade reports on the progress until 1977, and there is Looijenga's thorough introduction of 1984. On the following pages some essential features of this aspect and relations to the contents of the book will be sketched.

Let $f \in {}_{n+1}\mathfrak{m}$ have an isolated critical point. There are finitely many $h_1,\ldots,h_{r-1} \in {}_{n+1}\mathfrak{m}$ which represent a basis of ${}_{n+1}\mathfrak{m}/\langle \partial f/\partial z_1,\ldots,\partial f/\partial z_{n+1}, f\rangle$. The mapping germ $F: (\mathbb{C}^{n+1}\times\mathbb{C}^{r-1},0) \to (\mathbb{C}\times\mathbb{C}^{r-1},0)$ defined by

$$F(z,t) = (f(z)+t_1h_1(z)+\ldots+t_{r-1}h_{r-1}(z),\ t_1,\ldots,t_{r-1})$$

is called a *versal deformation* of the zero set $N(f)$. It is of course closely related to the universe unfolding of f, see §6, corollary 3.

Let $C = \{x\colon \operatorname{rk} D_xF < r\}$ denote the *critical locus* and $D = F(C)$ the *discriminant locus*. Let $\varepsilon>0$ be sufficiently small, let S be a suitable neighbourhood of 0 in $\mathbb{C}^r$. The restriction of F to $E = \{x\in\mathbb{C}^{n+r}\colon |x|\le\varepsilon,\ F(x)\in S\smallsetminus D\}$ is the projection of a locally trivial fibre bundle

$$p=f|E\colon E \to S\smallsetminus D\ ,$$

which is called the *Milnor fibration*. Fix a base point $s \in S\smallsetminus D$. The *Milnor fibre* $X=p^{-1}(s)$ is a compact oriented real $2n$-dimensional manifold which has the homotopy type of a bouquet of $\mu=\mu(f)$ many n-spheres, compare the remark in §2. Hence the only interesting homology group is $H_n(X)$. It is free of rank μ. The *intersection form* on $H_n(X)$ assigns to $u,v\in H_n(X)$ the intersection number $u\cdot v$, compare IV§11. It is symmetric for even n and skew for odd n. The fundamental group $\pi(S\smallsetminus D,s)$ acts on $H_n(X)$, preserving the intersection form. This action is called *monodromy*, and the image of $\pi(S\smallsetminus D,s)$ in the automorphism group of $H_n(X)$ is called the *monodromy group* Γ.

In $H_n(X)$ there is a distinguished Γ-invariant subset whose elements δ are called *vanishing cycles*. Their self-intersection number is $\delta\cdot\delta = (-1)^{n/2}2$ for even n and $=0$ for odd n. Every vanishing cycle determines the *pseudo-reflection* σ_δ in Γ given by

$$\sigma_\delta(v) = v-(-1)^{n(n-1)/2}(v\cdot\delta)\cdot\delta\ .$$

There is a basis of $H_n(X)$ consisting of vanishing cycles. The corresponding pseudo-reflections $\sigma_1,\ldots,\sigma_\mu$ generate Γ. The module $H_n(X)$ together with the subset of vanishing cycles is called the *vanishing lattice* V(f) of the germ f. Indeed, it depends up to isomorphism only on the germ of the zero set N(f) .

If f is adjacent to g , i.e. $g \prec f$ as defined in §1, V(g) is a sublattice of V(f) in such a way that a basis of V(g) extends to a basis of V(f) . For more properties of vanishing lattices we refer to Looijenga's book.

Let n be even. The intersection form on V(f) is definite if and only if f is simple. In this case, for n=2 , the Milnor fibre X is isomorphic to the minimal resolution Y of N(f) at 0 , more precisely diffeomorphic to a compact tubular neighbourhood of the exceptional set $E = E_1 \cup .. \cup E_k$ in Y , compare IV§10. The homology classes $[E_i]$ correspond to a basis of $H_2(X)$ consisting of vanishing cycles. The vanishing lattice V(f) for $f \in A_k, D_k, E_k$ is the simple root system of type A_k, D_k, E_k respectively, and the monodromy group is the corresponding Weyl group. For the notion of *root system* etc. see Bourbaki or Jacobson. In the context of root systems, the trees of the table in IV§10, which describe the intersection form (IV§12), are called *Dynkin diagrams* . All relevant information is encoded in these diagrams, e.g. the monodromy group Γ . It was this relationship which caused Arnold to denote the equivalence classes of simple germs by the traditional symbols A_k, D_k, E_k of the root systems. The adjacency relations between the simple germs, see §12, correspond exactly to the inclusions of the corresponding root systems, hence to the inclusions of the Dynkin diagrams.

Root systems are used in order to classify simple Lie groups. Brieskorn constructed directly the versal deformations of the simple singularities A_k, D_k, E_k by means of the corresponding Lie groups; for this see also Slodowy's book.
John McKay associated to the finite subgroups $G < SL(2,\mathbb{C})$ the Dynkin diagrams A_k, D_k, E_k by means of the representations of G ; see also Appendix III of Slodowy's book.

The finite subgroups of SO(3) correspond to the tesselations of the 2-sphere by congruent spherical triangles with angles

π/p , π/q , π/r where p, q, r are natural numbers. The angle sum restricts the possibilities: $1/p+1/q+1/r>1$. There are corresponding tesselations of the Euclidian plane, where $1/p + 1/q + 1/r = 1$ allows three possibilities, and of the hyperbolic plane, where $1/p + 1/q + 1/r < 1$ allows infinitely many possibilities. Dolgacev associated singularities with all these tesselations in such a way that the simple singularities belong to the spherical tesselations. The other singularities he obtains occur as next complicated ones after the simple singularities in the hierarchy; compare also Milnor's article.

There are still more possibilities to construct and to characterize simple singularities. Durfee compares 15 characterizations. Only some of them have been considered above.

References

V.I. Arnold, S.M. Gusein-Zade, A.N. Varchenko: Singularities of Differentiable Maps I. Birkhäuser Basel/Boston, 1985.

N. Bourbaki: Groupes et Algèbres de Lie. 5 vol. Hermann/Masson, Paris, 1962-1982.

E. Brieskorn: Rationale Singularitäten komplexer Flächen. Inventiones math. 4, 336-358 (1968).

E. Brieskorn: Singular Elements of Semisimple Algebraic Groups. Actes Congr. Intern. Mathém. Nice, vol. 2, 279-284 (1970).

A. Brønsted: An Introduction to Convex Polytopes. Springer, New York, 1983.

H. Cartan: Quotient d'un Espace Analytique par un Group d'Automorphismes. in: Algebraic Geometry and Topology. A symposium in honor of S. Lefschetz. Princeton Univ. Press, Princeton, 1957.

M.M. Cohen: A Course in Simple Homotopy Theory. Springer, New York, 1970.

H.S.M. Coxeter: Regular Polytopes. 3rd ed. Dover, New York, 1973.

H.S.M. Coxeter: Regular Complex Polytopes. Cambridge Univ. Press, Cambridge 1974.

A. Dold: Lectures on Algebraic Topology. Springer, Berlin 1972.

I.V. Dolgacev: Quotient-conic singularities of complex spaces. Funkt. Analiz i Ego Pril. 8:2,75-76 (1974), engl. transl.: Functional Analysis and its Appl. 8, 160-161 (1974).

I.V. Dolgacev: Automorphic forms and quasi-homogeneous singularities. Funkt. Analiz i Ego Pril. 9:2, 67-68 (1975), engl. transl.: Funktional Analysis and its Appl. 9, 149-150 (1975).

A. Durfee: Fifteen characterizations of rational double points and simple critical points. Enseignement Mathém. 25, 131-163 (1979).

P. DuVal: On isolated singularities which do not affect the condition of adjunction. Proc. Cambridge Phil. Soc. 30, 453-459 (1934).

P. DuVal: Homographies Quaternions and Rotations. Clarendon Press, Oxford, 1964.

Euclid: The Thirteen Books of the Elements. Edited by Th.L. Heath. Dover, New York, 1959.

W. Fischer, I. Lieb: Funktionentheorie. Vieweg, Braunschweig, 1981.

M. Golubitzki, V. Guillemin: Stable Mappings and Their Singularities. Springer, New York, 1973.

H. Grauert, R. Remmert: Coherent Analytic Sheaves. Springer, Berlin, 1984.

B. Grünbaum: Convex Polytopes. J. Wiley, New York, 1967.

R.C. Gunning: Lectures on Complex Analytic Varieties. The Local Parametrization Theorem. Finite Analytic Mappings. Princeton Univ. Press, Princeton 1970, 1974.

R.C. Gunning, H. Rossi: Analytic Functions of Several Complex Variables. Prentice Hall, Englewood Cliffs, 1965.

M.W. Hirsch: Differential Topology. Springer, New York, 1976.

F. Hirzebruch: Über vierdimensionale Riemannsche Flächen mehrdeutiger analytischer Funktionen von zwei komplexen Veränderlichen. Mathem. Annalen 126, 1-22 (1953).

S.M. Husein (or Gusein)-Zade: Monodromy groups of isolated singularities. Uspehi Mat. Nauk 32:2, 23-65 (1977), engl. transl. Russian Math. Surveys 32:2, 23-69 (1977).

N. Jacobson: Lie Algebras. Interscience, New York, 1962.

H.W.E. Jung: Darstellung der Funktionen eines algebraischen Körpers zweier unabhängigen Veränderlichen... Crelles J. für die reine und angewandte Mathem. 133, 289-314 (1908).

F. Klein: Vorlesungen über das Ikosaeder und die Auflösung der Gleichungen vom fünften Grade. Teubner, Leipzig, 1884.

E. Kunz: Einführung in die kommutative Algebra und algebraische Geometrie. Vieweg, Braunschweig, 1980.

S. Lang: Algebra. Addison-Wesley, Reading, 1965.

H.B. Laufer: Normal two-dimensional singularities. Ann. of Mathem. Studies 71, Princeton Univ. Press, Princeton, 1971.

E.J.N. Looijenga: Isolated Singular Points on Complete Intersections. London Math. Soc. Lect. Note Series 77. Cambridge Univ. Press, Cambridge, 1984.

J. McKay: Graphes, singularities, and finite groups. Proc. Symposia Pure Mathem. 37, 183-186 (1980).

W.S. Massey: Algebraic Topology: An Introduction. Harcourt, Brace & World, New York, 1967.

J. Milnor: Singular Points of Complex Hypersurfaces. Ann. of Mathem. Studies 61, Princeton Univ. Press, Princeton, 1968.

J. Milnor: On the 3-dimensional Brieskorn manifolds M(p,q,r). in: Knots, Groups and 3-manifolds. ed. by L.P. Neuwirth. Ann. of Mathem. Studies 84, Princeton Univ. Press, Princeton, 1975.

D. Mumford: The topology of normal singularities of an algebraic surface and a criterion for simplicity. Publ. Mathém. Inst. des Hautes Etudes Sci. 9, 229-246 (1961).

H. Pinkham: Singularités de Klein I, II in: Séminaire sur les Singularités des Surfaces. Ed. by M. Demazure, H. Pinkham, B. Teissier. Springer Lecture Notes in Mathem. 777, Berlin, 1980.

D. Prill: Local classification of quotients of complex manifolds by discontinuous groups. Duke Mathem. J. 34, 375-386 (1967).

H. Seifert, W. Threlfall: Topologische Untersuchung der Discontinuitätsbereiche endlicher Bewegungsgruppen des dreidimensionalen Raumes. Mathem. Annalen 104, 1-70 (1931).

H. Seifert, W. Threlfall: Lehrbuch der Topologie. Teubner, Leipzig, 1934.

I.R. Shafarevich: Basic Algebraic Geometry. Springer, Berlin, 1974.

P. Slodowy: Simple Singularities and Simple Algebraic Groups. Springer, Lecture Notes in Mathem. 815, Berlin, 1980.

O. Zariski, P. Samuel: Commutative Algebra, 2 vol. Van Nostrand, Princeton 1958, 1960.

Subject and Notation Index